This book belongs to

_____

Grade: _____

School: _____

Greetings, wonderful parents!

Thank you for choosing this amazing book to help your child learn **Fractions**. We hope you're just as excited as we are to begin this learning journey!

Your feedback is incredibly valuable to us. Please take a moment to leave a review on the platform where you purchased the book, and let us know what you thought. We're always striving to make our resources better and more effective, and your insights will help us do just that.

And, if you're ready for more learning adventures, check out our other books in the series. We promise they're just as fantastic as this one!

Thanks again for your support, and happy learning!

Best regards,
abcZbook Press
www.abczbook.com

A few minutes of math practice every day can help children master math skills. '**100 Days of Timed Tests: Fractions Practice**' is a beginner-level math practice workbook for Grade 4-5 kids. This book is specifically designed for different fraction problems such as finding fractions from a model, comparing two fractions, reducing or simplifying fractions, finding equivalent fractions, converting decimals to fractions, adding, and subtracting fractions. These sets of math practice worksheets are designed to test fraction problem-solving skills.

Kids can challenge themselves with timed test problems. This book mainly focuses on improving math skills, and speed, and building confidence levels. The book also has an answer key sheet at the end of the book so that you can quickly check the kid's answer. In this book, there is a set of problems to be solved on a daily basis, and a total of 100 pages of timed test practice worksheets. It helps kids perform consistently and be trained to excel in fractions.

abcZbook Press

Table of Contents

No.	Math Problems Description	Days	☑
1	Finding fractions from a model	1 – 10	☐
2	Comparing the two fractions	11 – 20	☐
3	Reducing fractions to the simplest form	21 – 30	☐
4	Equivalent fractions	31 – 40	☐
5	Converting decimals to fractions	41 – 50	☐
6	Adding fractions	51 – 75	☐
7	Subtracting fractions	76 – 100	☐
8	Answer Key Sheets	-	☐
9	Fractions Chart	-	☐
10	Certificate of Achievement	☆☆☆☆☆	☐

Find the fraction from each model:

(1) $= \dfrac{}{}$

(2) $= \dfrac{}{}$

(3) $= \dfrac{}{}$

(4) $= \dfrac{}{}$

(5) $= \dfrac{}{}$

(6) $= \dfrac{}{}$

(7) $= \dfrac{}{}$

(8) $= \dfrac{}{}$

(9) $= \dfrac{}{}$

(10) $= \dfrac{}{}$

(11) $= \dfrac{}{}$

(12) $= \dfrac{}{}$

(13) $= \dfrac{}{}$

(14) $= \dfrac{}{}$

(15) $= \dfrac{}{}$

(16) $= \dfrac{}{}$

(17) $= \dfrac{}{}$

(18) $= \dfrac{}{}$

Day:	2	Date:		Score:	/18
Name:		Time:	:	Rating:	☆☆☆☆☆

Find the fraction from each model:

(1)
= —

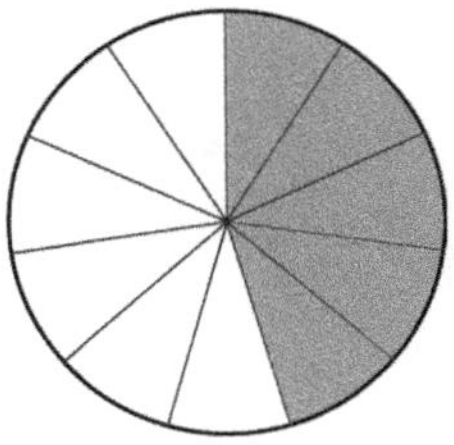

(2)
= —

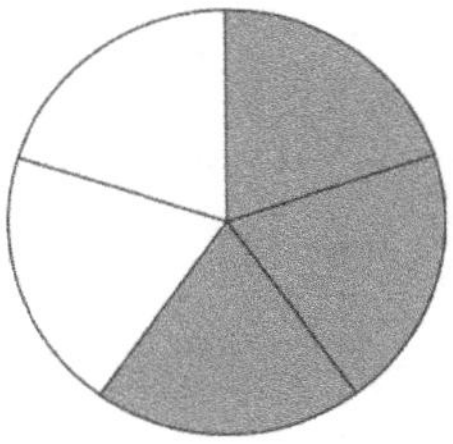

(3)
= —

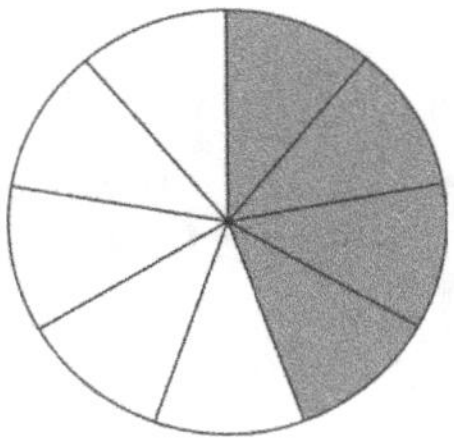

(4)
= —

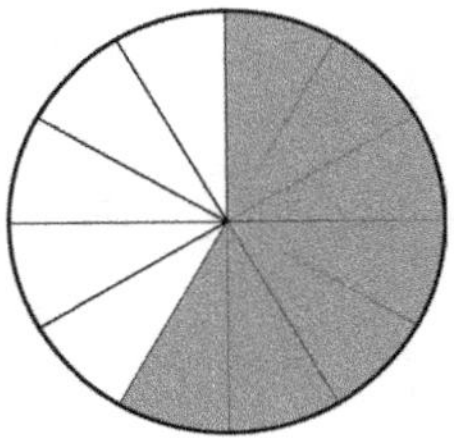

(5)
= —

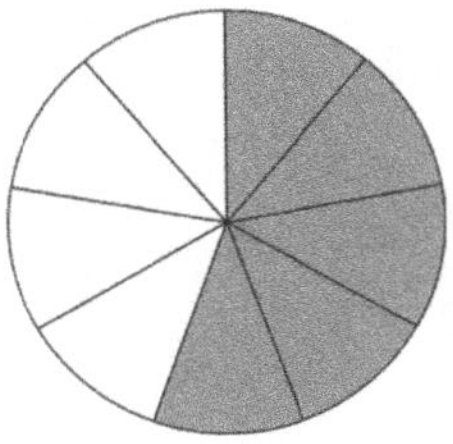

(6)
= —

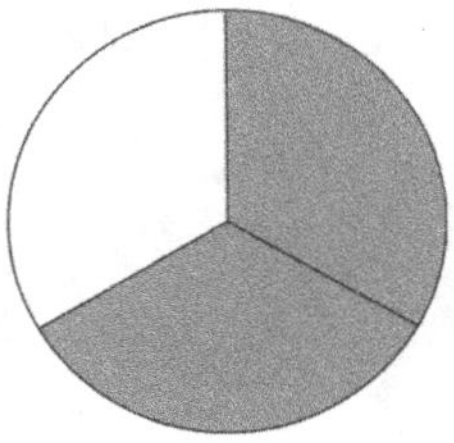

(7)
= —

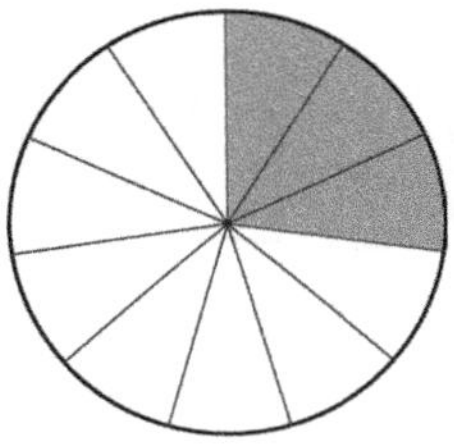

(8)
= —

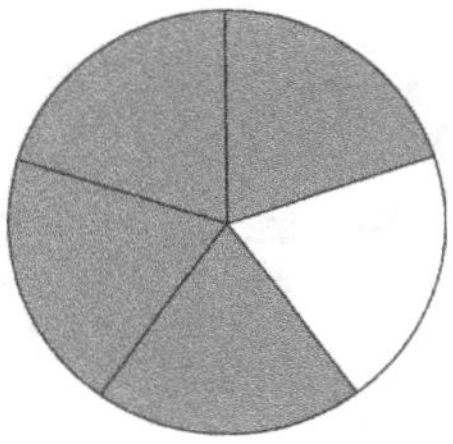

(9)
= —

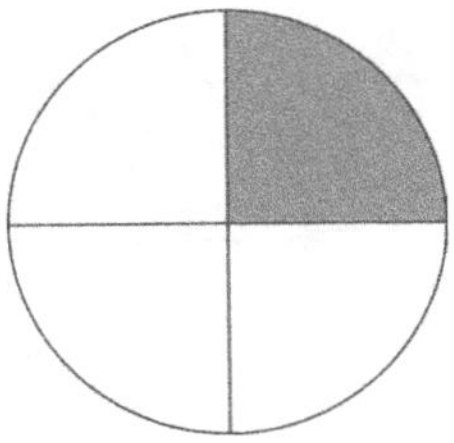

(10)
= —

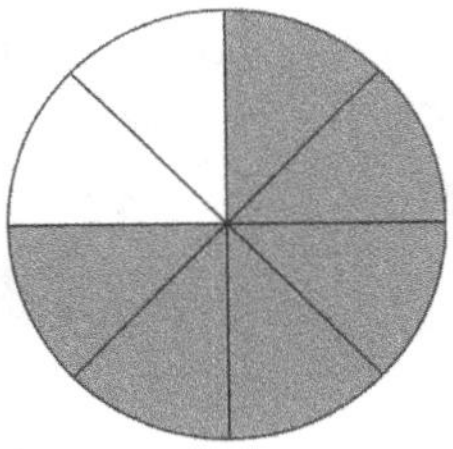

(11)
= —

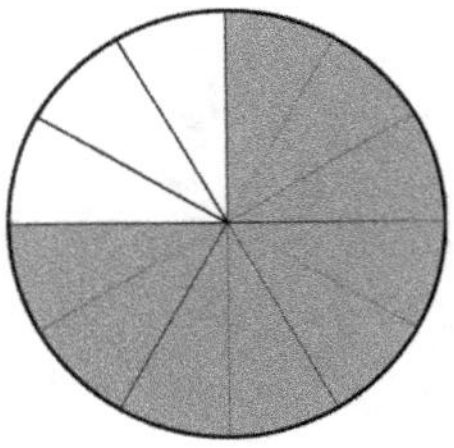

(12)
= —

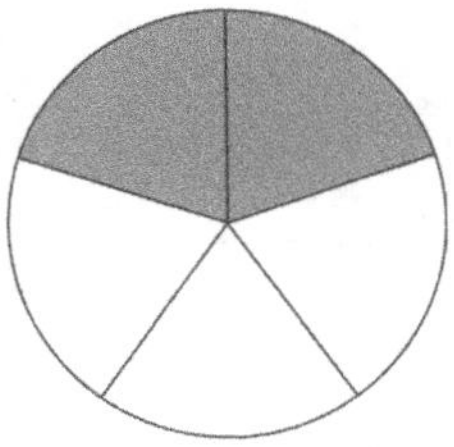

(13)
= —

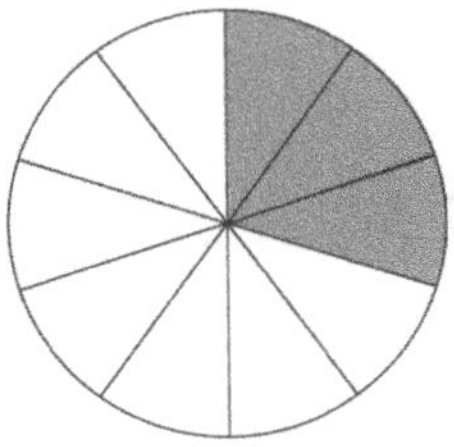

(14)
= —

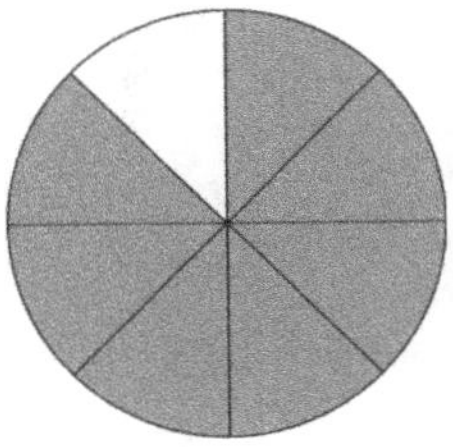

(15)
= —

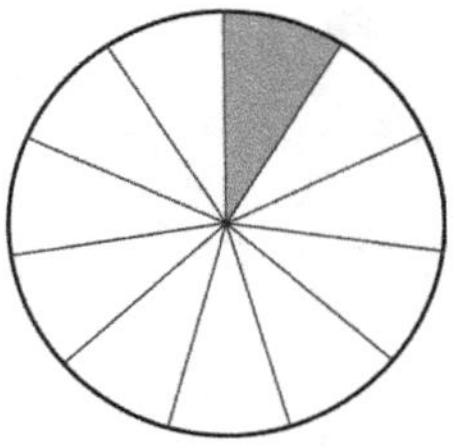

(16)
= —

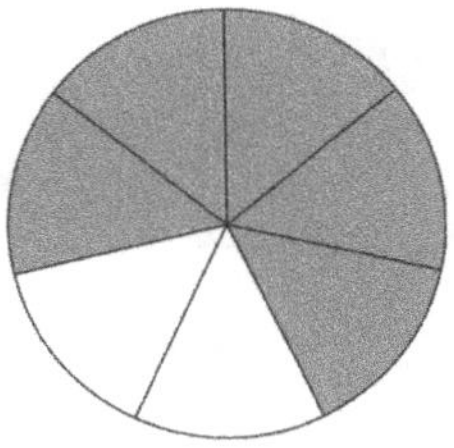

(17)
= —

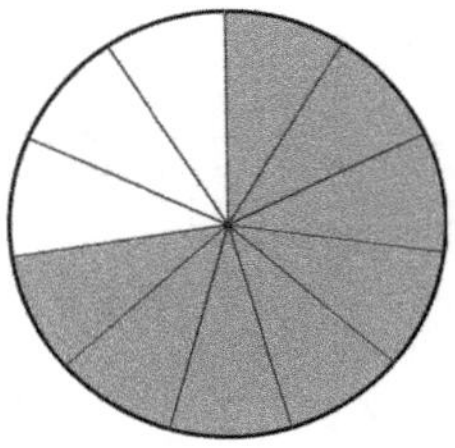

(18)
= —

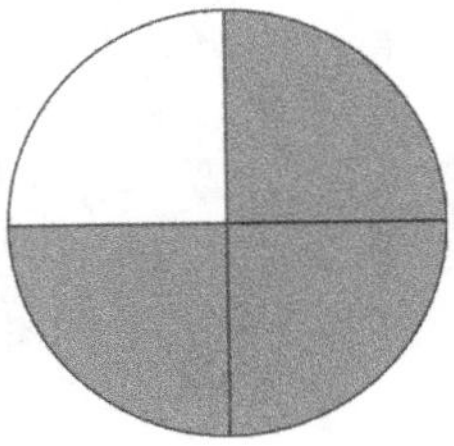

Find the fraction from each model:

(1) = —

(2) = —

(3) = —

(4) = —

(5) = —

(6) = —

(7) = —

(8) = —

(9) = —

(10) = —

(11) = —

(12) = —

(13) = —

(14) = —

(15) = —

(16) = —

(17) = —

(18) = —

Day:	4	Date:		Score:	/18
Name:		Time:	:	Rating:	☆☆☆☆☆

Find the fraction from each model:

(1) 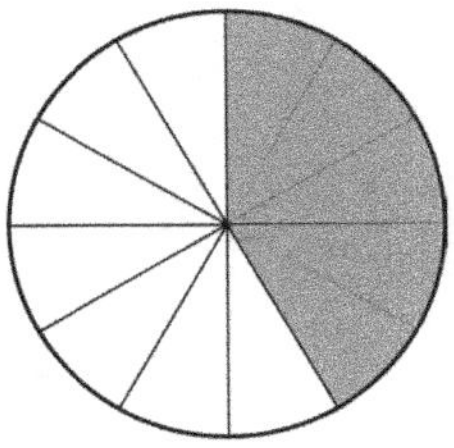 = —

(2) 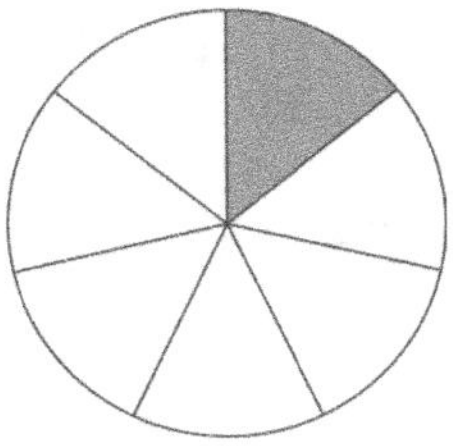 = —

(3) 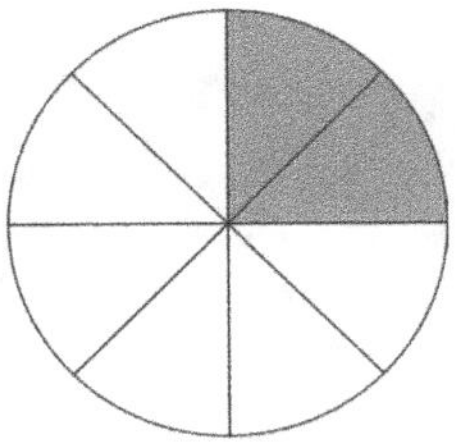 = —

(4) 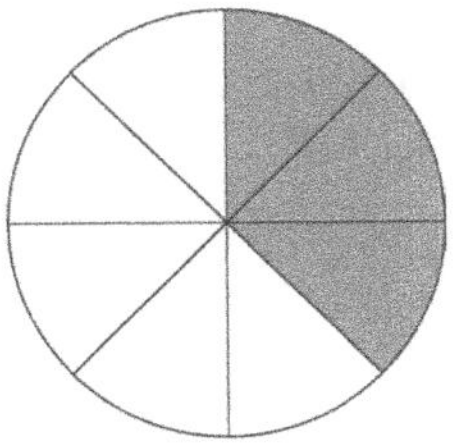 = —

(5) 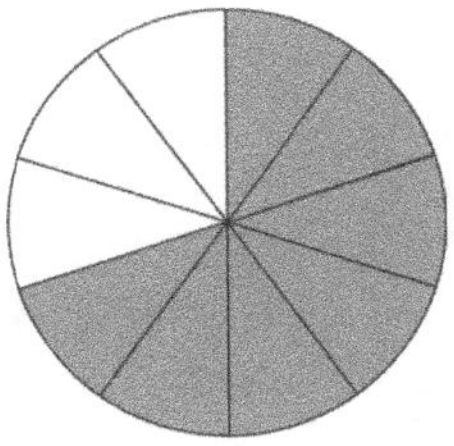 = —

(6) 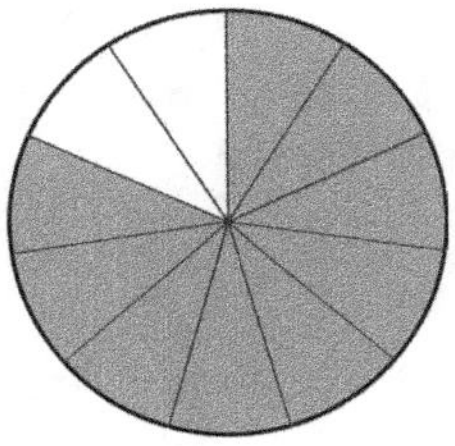 = —

(7) 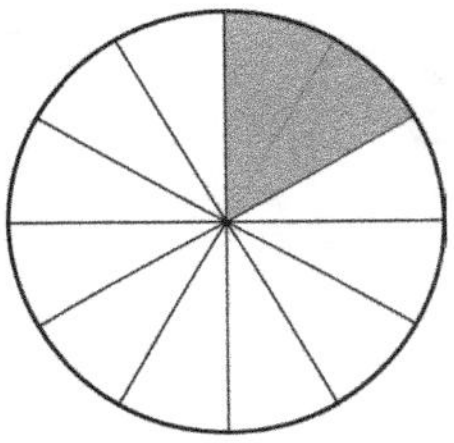 = —

(8) 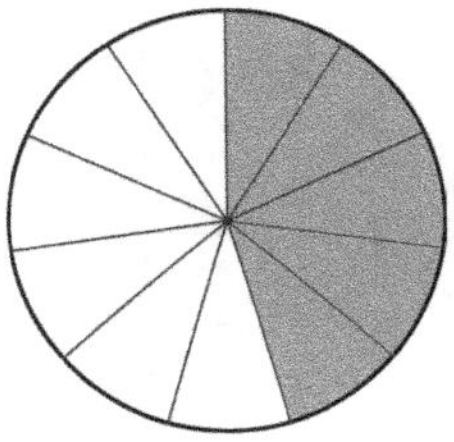 = —

(9) 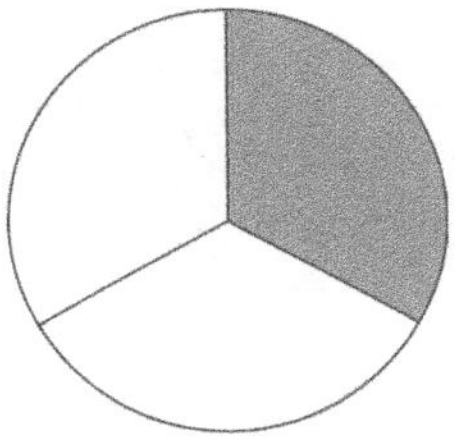 = —

(10) 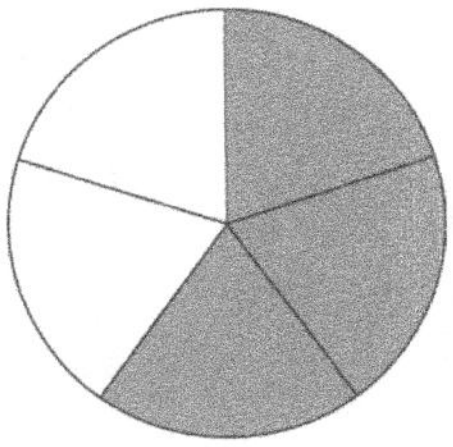 = —

(11) 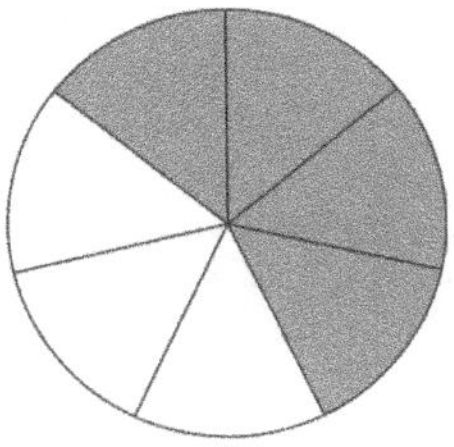 = —

(12) 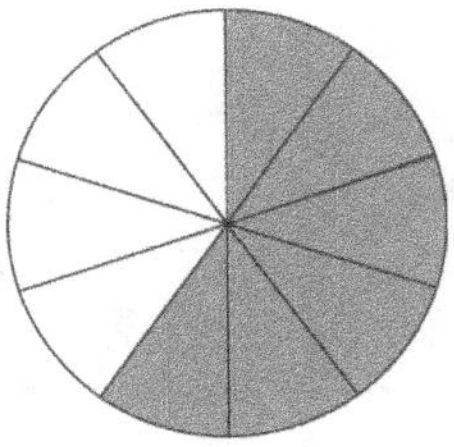 = —

(13) 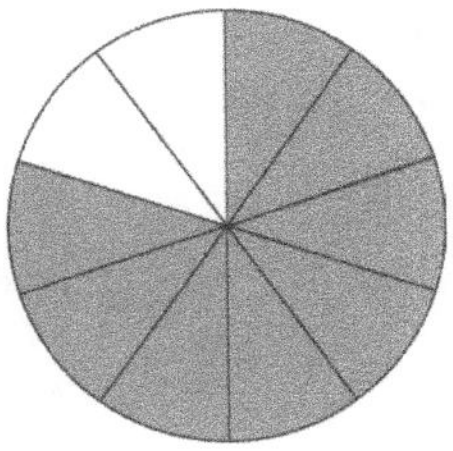 = —

(14) 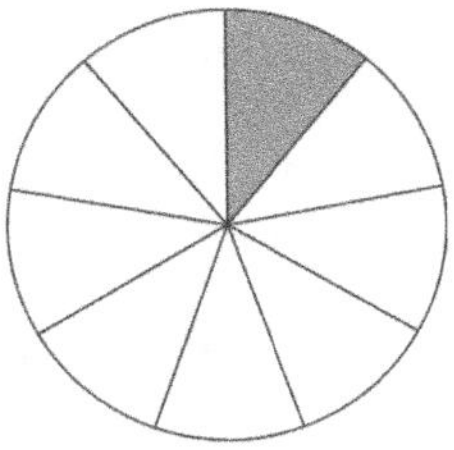 = —

(15) 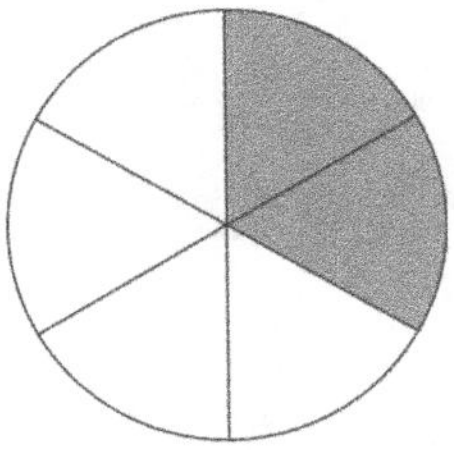 = —

(16) 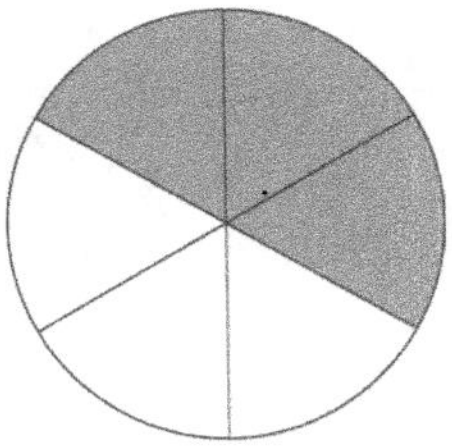 = —

(17) 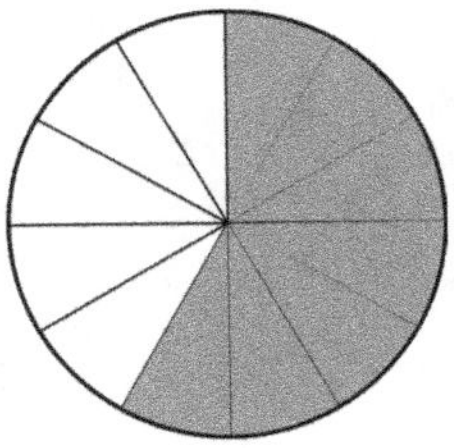 = —

(18) 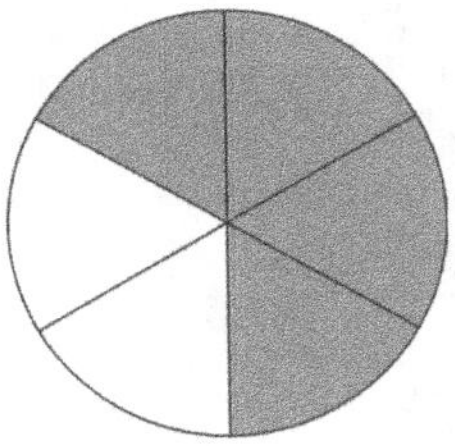 = —

Find the fraction from each model:

(1) $= \dfrac{\quad}{\quad}$

(2) $= \dfrac{\quad}{\quad}$

(3) $= \dfrac{\quad}{\quad}$

(4) $= \dfrac{\quad}{\quad}$

(5) $= \dfrac{\quad}{\quad}$

(6) $= \dfrac{\quad}{\quad}$

(7) $= \dfrac{\quad}{\quad}$

(8) $= \dfrac{\quad}{\quad}$

(9) $= \dfrac{\quad}{\quad}$

(10) $= \dfrac{\quad}{\quad}$

(11) $= \dfrac{\quad}{\quad}$

(12) $= \dfrac{\quad}{\quad}$

(13) $= \dfrac{\quad}{\quad}$

(14) $= \dfrac{\quad}{\quad}$

(15) $= \dfrac{\quad}{\quad}$

(16) $= \dfrac{\quad}{\quad}$

(17) $= \dfrac{\quad}{\quad}$

(18) $= \dfrac{\quad}{\quad}$

Find the fraction from each model:

(1) $= \dfrac{\quad}{\quad}$

(2) $= \dfrac{\quad}{\quad}$

(3) $= \dfrac{\quad}{\quad}$

(4) $= \dfrac{\quad}{\quad}$

(5) $= \dfrac{\quad}{\quad}$

(6) $= \dfrac{\quad}{\quad}$

(7) $= \dfrac{\quad}{\quad}$

(8) $= \dfrac{\quad}{\quad}$

(9) $= \dfrac{\quad}{\quad}$

(10) $= \dfrac{\quad}{\quad}$

(11) $= \dfrac{\quad}{\quad}$

(12) $= \dfrac{\quad}{\quad}$

(13) $= \dfrac{\quad}{\quad}$

(14) $= \dfrac{\quad}{\quad}$

(15) $= \dfrac{\quad}{\quad}$

(16) $= \dfrac{\quad}{\quad}$

(17) $= \dfrac{\quad}{\quad}$

(18) $= \dfrac{\quad}{\quad}$

Day: 7 Date: Score: /18

Name: Time: : Rating: ☆☆☆☆☆

Find the fraction from each model:

(1)

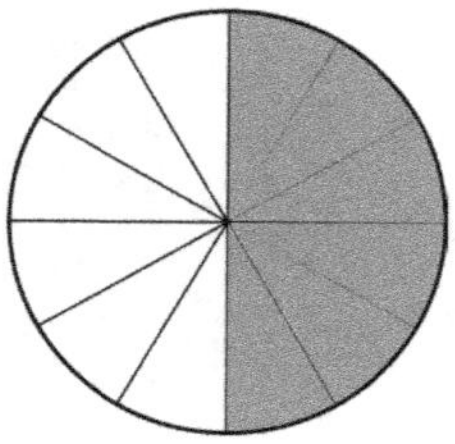

= —

(2)

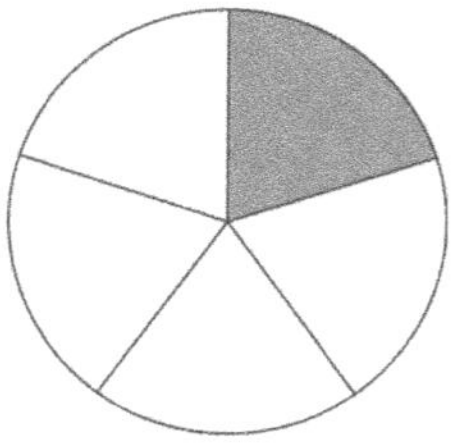

= —

(3)

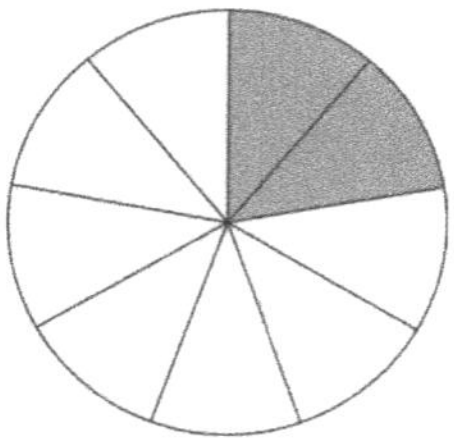

= —

(4)

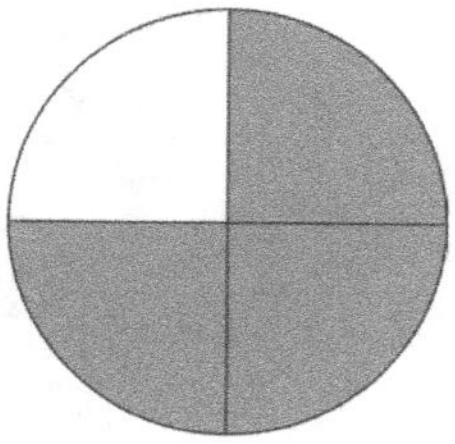

= —

(5)

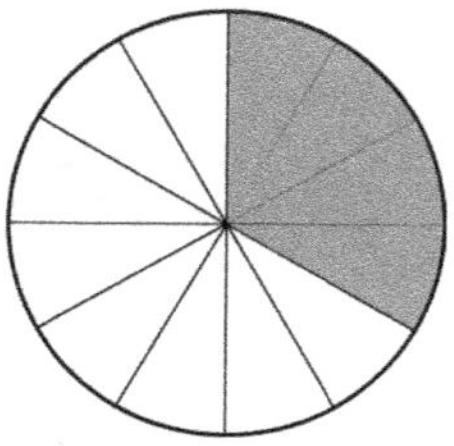

= —

(6)

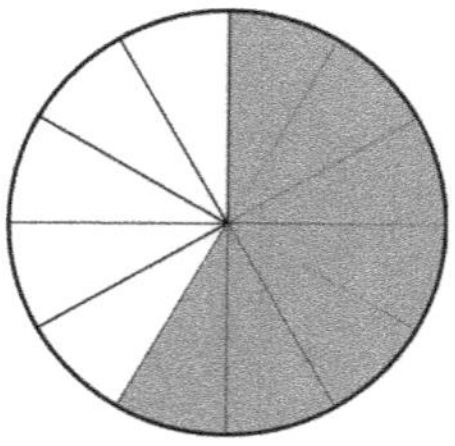

= —

(7)

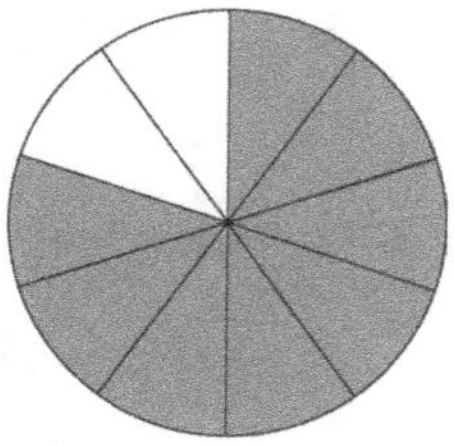

= —

(8)

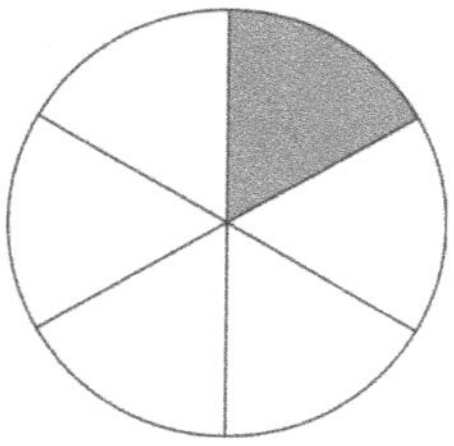

= —

(9)

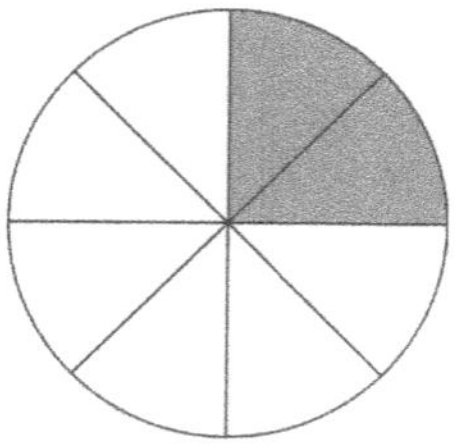

= —

(10)

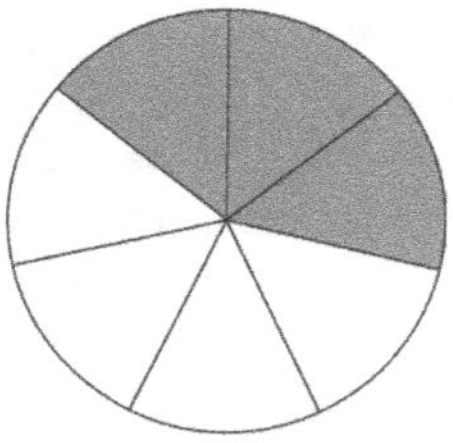

= —

(11)

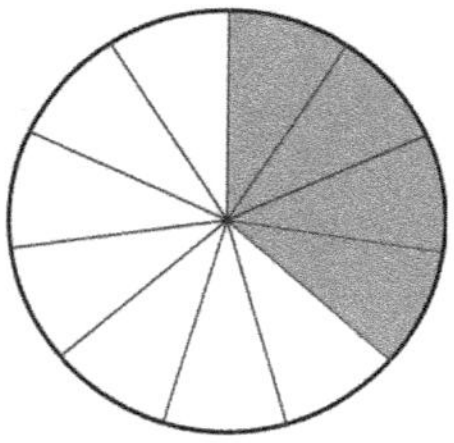

= —

(12)

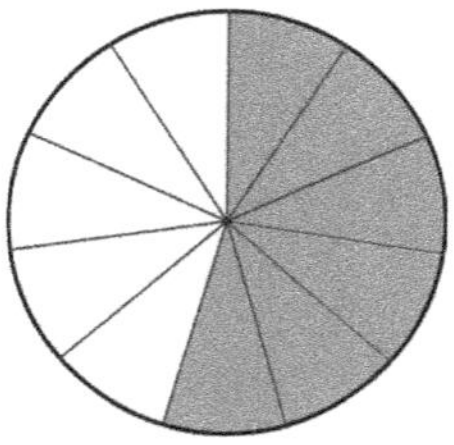

= —

(13)

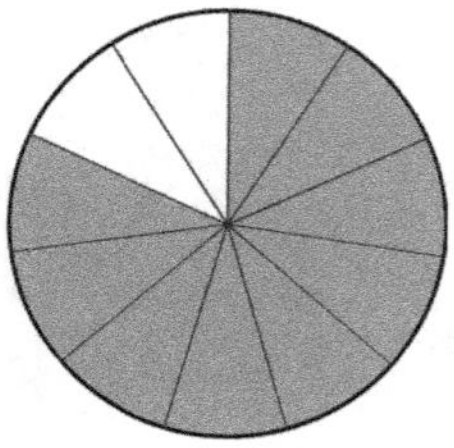

= —

(14)

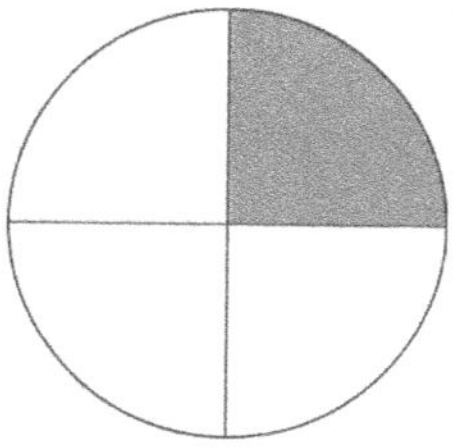

= —

(15)

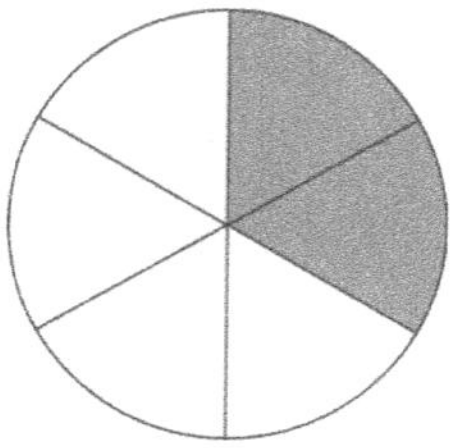

= —

(16)

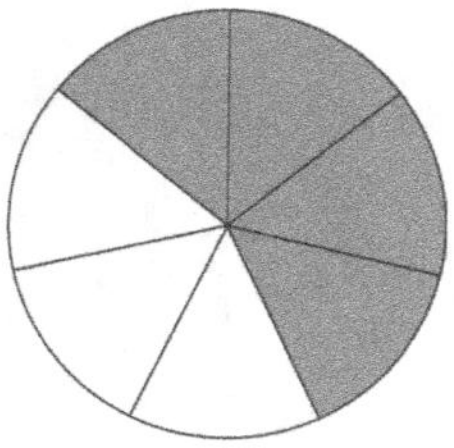

= —

(17)

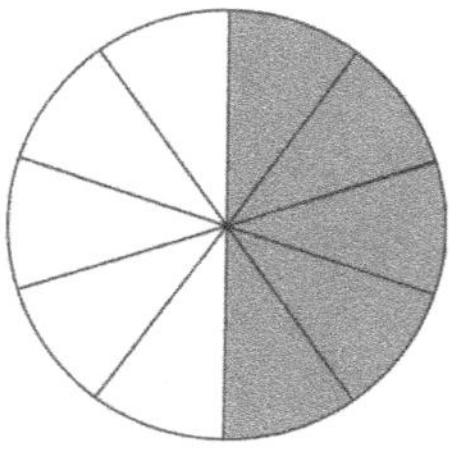

= —

(18)

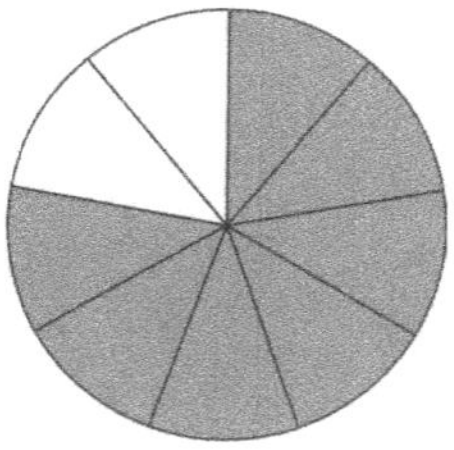

= —

Find the fraction from each model:

(1) $= \dfrac{\quad}{\quad}$

(2) $= \dfrac{\quad}{\quad}$

(3) $= \dfrac{\quad}{\quad}$

(4) $= \dfrac{\quad}{\quad}$

(5) $= \dfrac{\quad}{\quad}$

(6) $= \dfrac{\quad}{\quad}$

(7) $= \dfrac{\quad}{\quad}$

(8) $= \dfrac{\quad}{\quad}$

(9) $= \dfrac{\quad}{\quad}$

(10) $= \dfrac{\quad}{\quad}$

(11) $= \dfrac{\quad}{\quad}$

(12) $= \dfrac{\quad}{\quad}$

(13) $= \dfrac{\quad}{\quad}$

(14) $= \dfrac{\quad}{\quad}$

(15) $= \dfrac{\quad}{\quad}$

(16) $= \dfrac{\quad}{\quad}$

(17) $= \dfrac{\quad}{\quad}$

(18) $= \dfrac{\quad}{\quad}$

Day: 9 | Date: | Score: /18

Name: | Time: : | Rating: ☆☆☆☆☆

Find the fraction from each model:

(1)
= —

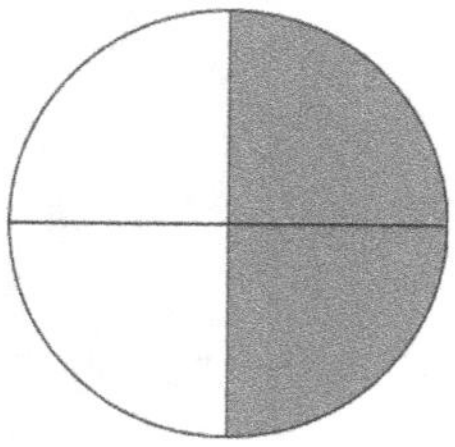

(2)
= —

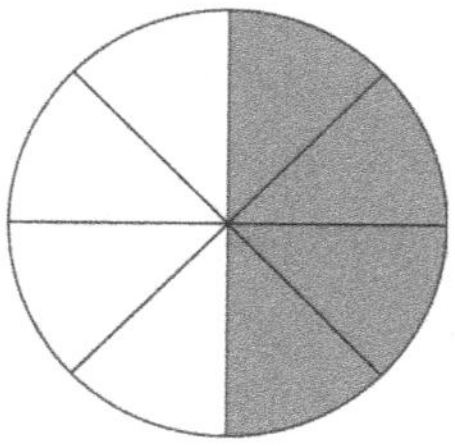

(3)
= —

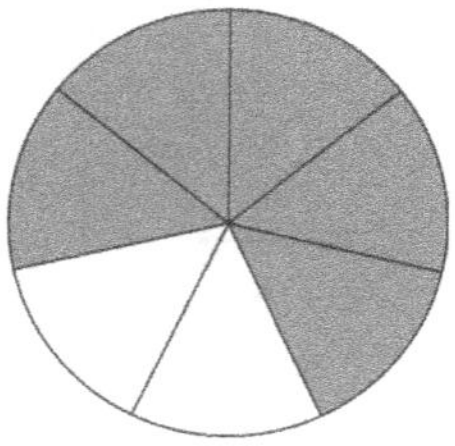

(4)
= —

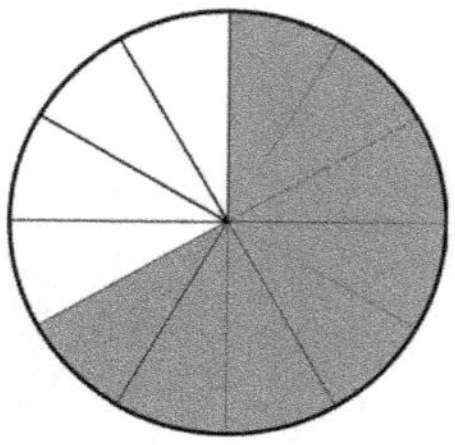

(5)
= —

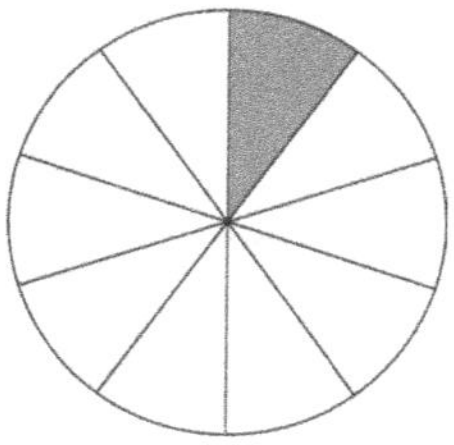

(6)
= —

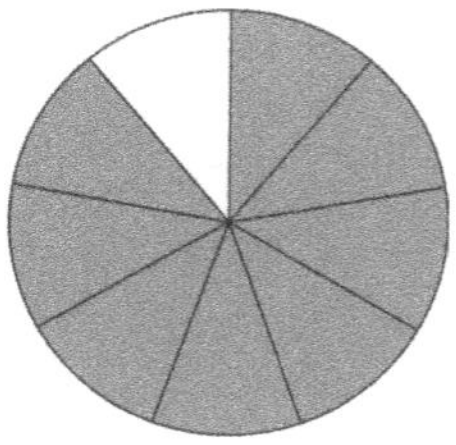

(7)
= —

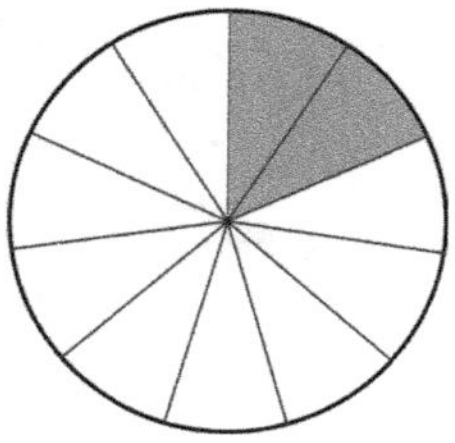

(8)
= —

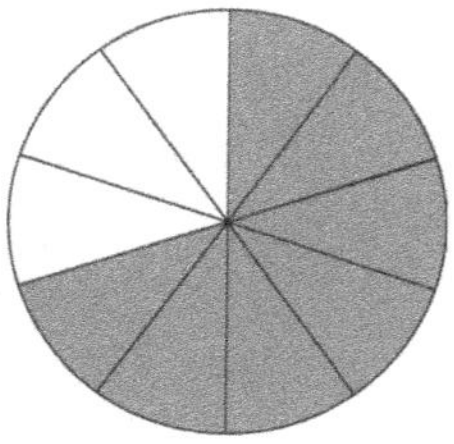

(9)
= —

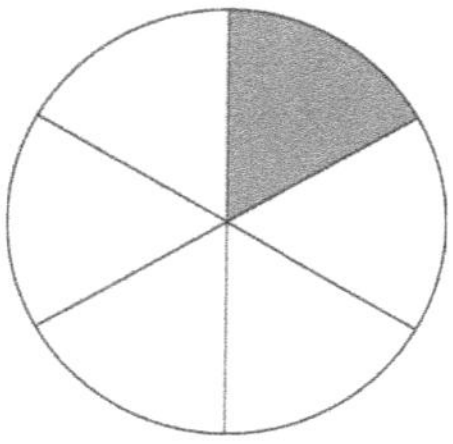

(10)
= —

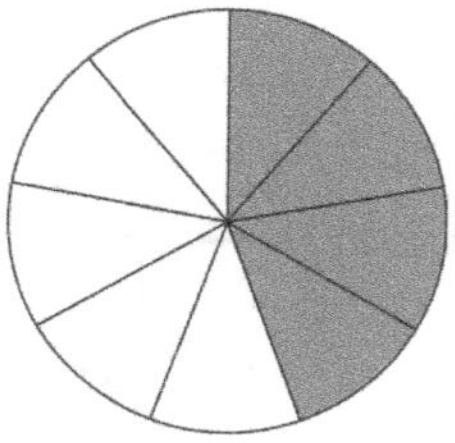

(11)
= —

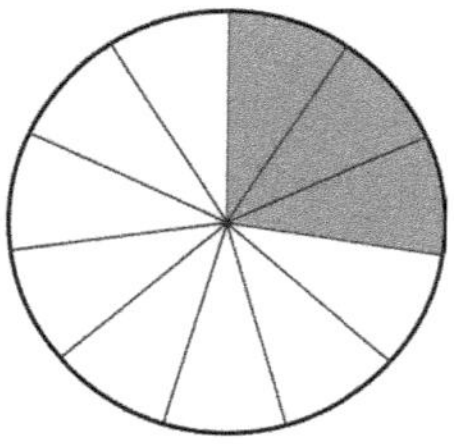

(12)
= —

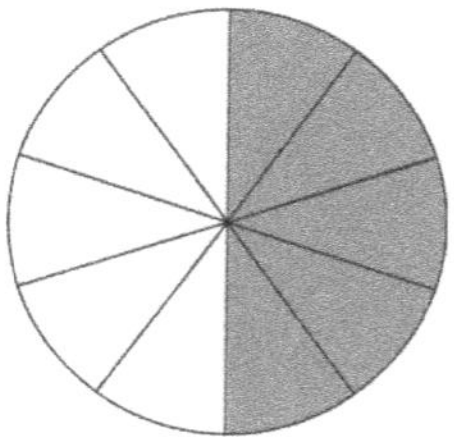

(13)
= —

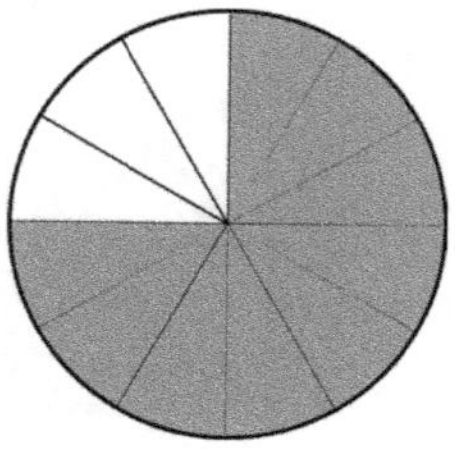

(14)
= —

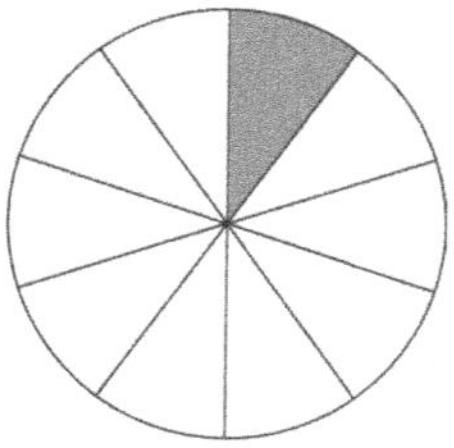

(15)
= —

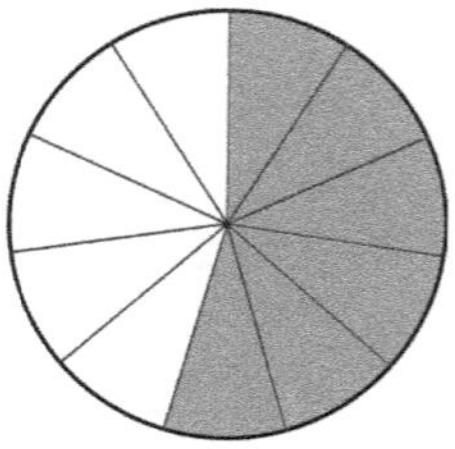

(16)
= —

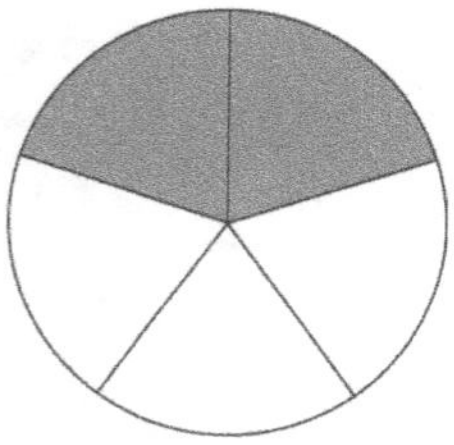

(17)
= —

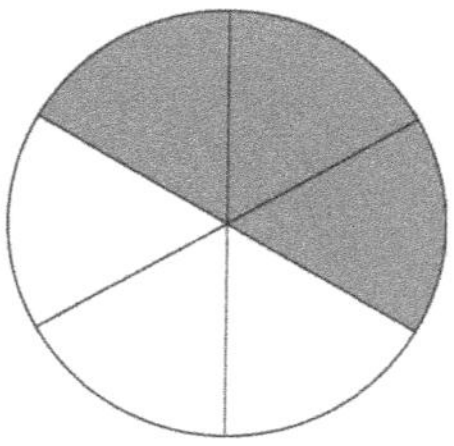

(18)
= —

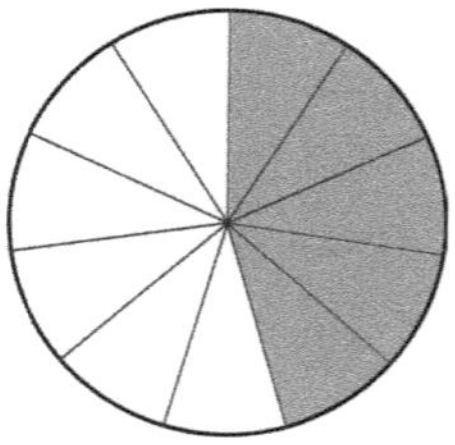

Find the fraction from each model:

(1)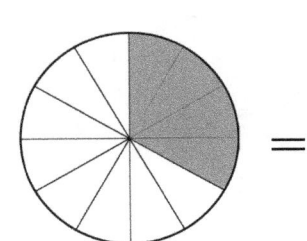

(2)

(3)

$= \underline{\quad}$

(4)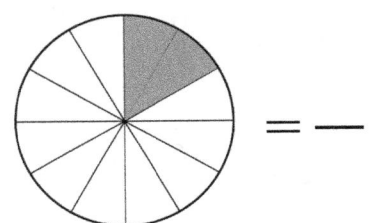

(5)

(6)

$= \underline{\quad}$

(7)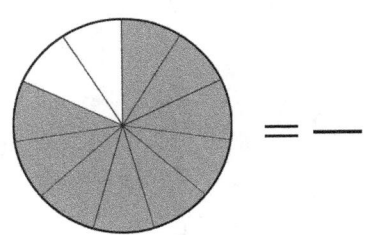

(8)

(9)

$= \underline{\quad}$

(10)

(11)

(12)

$= \underline{\quad}$

(13)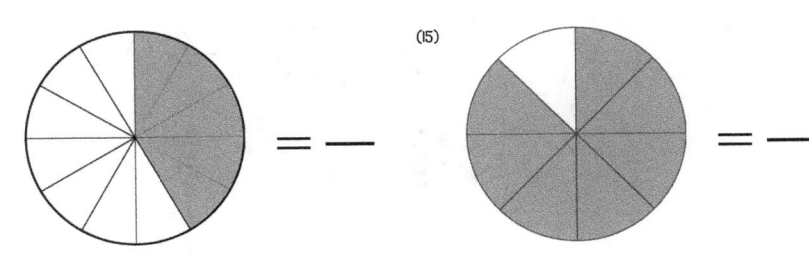

(14)

(15)

$= \underline{\quad}$

(16)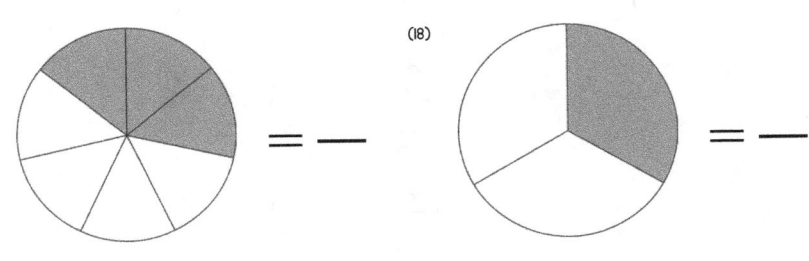

(17)

(18)

$= \underline{\quad}$

Comparing the two fractions by showing >, <, = symbol:

(1) $\dfrac{5}{6}$ ◯ $\dfrac{4}{6}$ (2) $\dfrac{2}{6}$ ◯ $\dfrac{1}{6}$ (3) $\dfrac{1}{3}$ ◯ $\dfrac{2}{3}$

(4) $\dfrac{2}{4}$ ◯ $\dfrac{3}{4}$ (5) $\dfrac{5}{9}$ ◯ $\dfrac{4}{9}$ (6) $\dfrac{2}{5}$ ◯ $\dfrac{1}{5}$

(7) $\dfrac{7}{8}$ ◯ $\dfrac{2}{8}$ (8) $\dfrac{4}{12}$ ◯ $\dfrac{7}{12}$ (9) $\dfrac{1}{6}$ ◯ $\dfrac{5}{6}$

(10) $\dfrac{5}{9}$ ◯ $\dfrac{1}{9}$ (11) $\dfrac{2}{3}$ ◯ $\dfrac{1}{3}$ (12) $\dfrac{3}{4}$ ◯ $\dfrac{1}{4}$

(13) $\dfrac{3}{10}$ ◯ $\dfrac{5}{10}$ (14) $\dfrac{1}{2}$ ◯ $\dfrac{1}{2}$ (15) $\dfrac{2}{7}$ ◯ $\dfrac{5}{7}$

(16) $\dfrac{1}{5}$ ◯ $\dfrac{3}{5}$ (17) $\dfrac{7}{10}$ ◯ $\dfrac{4}{10}$ (18) $\dfrac{4}{9}$ ◯ $\dfrac{1}{9}$

(19) $\dfrac{2}{6}$ ◯ $\dfrac{4}{6}$ (20) $\dfrac{5}{8}$ ◯ $\dfrac{2}{8}$ (21) $\dfrac{3}{8}$ ◯ $\dfrac{1}{8}$

(22) $\dfrac{5}{10}$ ◯ $\dfrac{2}{10}$ (23) $\dfrac{2}{7}$ ◯ $\dfrac{5}{7}$ (24) $\dfrac{2}{10}$ ◯ $\dfrac{3}{10}$

(25) $\dfrac{1}{2}$ ◯ $\dfrac{1}{2}$ (26) $\dfrac{3}{4}$ ◯ $\dfrac{1}{4}$ (27) $\dfrac{5}{12}$ ◯ $\dfrac{2}{12}$

(28) $\dfrac{4}{12}$ ◯ $\dfrac{7}{12}$ (29) $\dfrac{2}{5}$ ◯ $\dfrac{3}{5}$ (30) $\dfrac{4}{10}$ ◯ $\dfrac{2}{10}$

(31) $\dfrac{3}{7}$ ◯ $\dfrac{2}{7}$ (32) $\dfrac{4}{7}$ ◯ $\dfrac{2}{7}$ (33) $\dfrac{1}{9}$ ◯ $\dfrac{4}{9}$

(34) $\dfrac{1}{6}$ ◯ $\dfrac{3}{6}$ (35) $\dfrac{3}{8}$ ◯ $\dfrac{5}{8}$ (36) $\dfrac{2}{8}$ ◯ $\dfrac{6}{8}$

(37) $\dfrac{5}{10}$ ◯ $\dfrac{3}{10}$ (38) $\dfrac{1}{8}$ ◯ $\dfrac{1}{8}$ (39) $\dfrac{1}{10}$ ◯ $\dfrac{3}{10}$

(40) $\dfrac{2}{3}$ ◯ $\dfrac{1}{3}$ (41) $\dfrac{3}{12}$ ◯ $\dfrac{3}{12}$ (42) $\dfrac{2}{6}$ ◯ $\dfrac{3}{6}$

(43) $\dfrac{7}{9}$ ◯ $\dfrac{4}{9}$ (44) $\dfrac{2}{3}$ ◯ $\dfrac{2}{3}$ (45) $\dfrac{1}{7}$ ◯ $\dfrac{3}{7}$

Comparing the two fractions by showing >, <, = symbol:

(1) $\dfrac{4}{8}$ ◯ $\dfrac{3}{8}$ (2) $\dfrac{3}{8}$ ◯ $\dfrac{2}{8}$ (3) $\dfrac{3}{10}$ ◯ $\dfrac{7}{10}$

(4) $\dfrac{2}{10}$ ◯ $\dfrac{4}{10}$ (5) $\dfrac{5}{6}$ ◯ $\dfrac{1}{6}$ (6) $\dfrac{2}{5}$ ◯ $\dfrac{3}{5}$

(7) $\dfrac{3}{4}$ ◯ $\dfrac{1}{4}$ (8) $\dfrac{1}{8}$ ◯ $\dfrac{5}{8}$ (9) $\dfrac{3}{12}$ ◯ $\dfrac{7}{12}$

(10) $\dfrac{1}{8}$ ◯ $\dfrac{3}{8}$ (11) $\dfrac{7}{9}$ ◯ $\dfrac{5}{9}$ (12) $\dfrac{4}{7}$ ◯ $\dfrac{3}{7}$

(13) $\dfrac{6}{10}$ ◯ $\dfrac{2}{10}$ (14) $\dfrac{3}{5}$ ◯ $\dfrac{2}{5}$ (15) $\dfrac{5}{10}$ ◯ $\dfrac{6}{10}$

(16) $\dfrac{2}{7}$ ◯ $\dfrac{1}{7}$ (17) $\dfrac{2}{6}$ ◯ $\dfrac{5}{6}$ (18) $\dfrac{1}{8}$ ◯ $\dfrac{6}{8}$

(19) $\dfrac{4}{6}$ ◯ $\dfrac{2}{6}$ (20) $\dfrac{1}{3}$ ◯ $\dfrac{2}{3}$ (21) $\dfrac{2}{9}$ ◯ $\dfrac{5}{9}$

(22) $\dfrac{5}{12}$ ◯ $\dfrac{4}{12}$ (23) $\dfrac{4}{7}$ ◯ $\dfrac{4}{7}$ (24) $\dfrac{2}{7}$ ◯ $\dfrac{5}{7}$

(25) $\dfrac{1}{9}$ ◯ $\dfrac{2}{9}$ (26) $\dfrac{3}{4}$ ◯ $\dfrac{1}{4}$ (27) $\dfrac{4}{5}$ ◯ $\dfrac{1}{5}$

(28) $\dfrac{3}{5}$ ◯ $\dfrac{2}{5}$ (29) $\dfrac{5}{8}$ ◯ $\dfrac{2}{8}$ (30) $\dfrac{3}{8}$ ◯ $\dfrac{4}{8}$

(31) $\dfrac{6}{7}$ ◯ $\dfrac{1}{7}$ (32) $\dfrac{2}{4}$ ◯ $\dfrac{4}{4}$ (33) $\dfrac{4}{10}$ ◯ $\dfrac{7}{10}$

(34) $\dfrac{2}{8}$ ◯ $\dfrac{1}{8}$ (35) $\dfrac{1}{9}$ ◯ $\dfrac{4}{9}$ (36) $\dfrac{1}{10}$ ◯ $\dfrac{3}{10}$

(37) $\dfrac{4}{10}$ ◯ $\dfrac{6}{10}$ (38) $\dfrac{4}{10}$ ◯ $\dfrac{2}{10}$ (39) $\dfrac{5}{6}$ ◯ $\dfrac{2}{6}$

(40) $\dfrac{1}{4}$ ◯ $\dfrac{3}{4}$ (41) $\dfrac{3}{8}$ ◯ $\dfrac{4}{8}$ (42) $\dfrac{4}{9}$ ◯ $\dfrac{5}{9}$

(43) $\dfrac{5}{9}$ ◯ $\dfrac{3}{9}$ (44) $\dfrac{1}{6}$ ◯ $\dfrac{3}{6}$ (45) $\dfrac{1}{5}$ ◯ $\dfrac{4}{5}$

Comparing the two fractions by showing >, <, = symbol:

(1) $\dfrac{2}{10}$ ◯ $\dfrac{1}{10}$ (2) $\dfrac{2}{8}$ ◯ $\dfrac{4}{8}$ (3) $\dfrac{3}{10}$ ◯ $\dfrac{4}{10}$

(4) $\dfrac{3}{6}$ ◯ $\dfrac{2}{6}$ (5) $\dfrac{3}{6}$ ◯ $\dfrac{3}{6}$ (6) $\dfrac{2}{10}$ ◯ $\dfrac{3}{10}$

(7) $\dfrac{4}{5}$ ◯ $\dfrac{1}{5}$ (8) $\dfrac{1}{7}$ ◯ $\dfrac{1}{7}$ (9) $\dfrac{3}{7}$ ◯ $\dfrac{2}{7}$

(10) $\dfrac{7}{12}$ ◯ $\dfrac{4}{12}$ (11) $\dfrac{2}{5}$ ◯ $\dfrac{2}{5}$ (12) $\dfrac{1}{12}$ ◯ $\dfrac{3}{12}$

(13) $\dfrac{1}{7}$ ◯ $\dfrac{4}{7}$ (14) $\dfrac{4}{9}$ ◯ $\dfrac{2}{9}$ (15) $\dfrac{4}{10}$ ◯ $\dfrac{3}{10}$

(16) $\dfrac{2}{9}$ ◯ $\dfrac{4}{9}$ (17) $\dfrac{1}{5}$ ◯ $\dfrac{4}{5}$ (18) $\dfrac{5}{8}$ ◯ $\dfrac{1}{8}$

(19) $\dfrac{3}{10}$ ◯ $\dfrac{2}{10}$ (20) $\dfrac{5}{7}$ ◯ $\dfrac{2}{7}$ (21) $\dfrac{2}{12}$ ◯ $\dfrac{5}{12}$

(22) $\dfrac{4}{10}$ ◯ $\dfrac{3}{10}$ (23) $\dfrac{3}{10}$ ◯ $\dfrac{3}{10}$ (24) $\dfrac{2}{4}$ ◯ $\dfrac{1}{4}$

(25) $\dfrac{5}{7}$ ◯ $\dfrac{3}{7}$ (26) $\dfrac{2}{12}$ ◯ $\dfrac{4}{12}$ (27) $\dfrac{1}{10}$ ◯ $\dfrac{2}{10}$

(28) $\dfrac{1}{3}$ ◯ $\dfrac{2}{3}$ (29) $\dfrac{1}{10}$ ◯ $\dfrac{3}{10}$ (30) $\dfrac{3}{6}$ ◯ $\dfrac{2}{6}$

(31) $\dfrac{2}{4}$ ◯ $\dfrac{1}{4}$ (32) $\dfrac{4}{6}$ ◯ $\dfrac{2}{6}$ (33) $\dfrac{2}{3}$ ◯ $\dfrac{1}{3}$

(34) $\dfrac{3}{8}$ ◯ $\dfrac{2}{8}$ (35) $\dfrac{3}{7}$ ◯ $\dfrac{1}{7}$ (36) $\dfrac{1}{4}$ ◯ $\dfrac{3}{4}$

(37) $\dfrac{4}{7}$ ◯ $\dfrac{1}{7}$ (38) $\dfrac{2}{9}$ ◯ $\dfrac{2}{9}$ (39) $\dfrac{4}{8}$ ◯ $\dfrac{6}{8}$

(40) $\dfrac{5}{10}$ ◯ $\dfrac{13}{10}$ (41) $\dfrac{1}{12}$ ◯ $\dfrac{3}{12}$ (42) $\dfrac{5}{9}$ ◯ $\dfrac{3}{9}$

(43) $\dfrac{1}{5}$ ◯ $\dfrac{2}{5}$ (44) $\dfrac{5}{8}$ ◯ $\dfrac{3}{8}$ (45) $\dfrac{3}{5}$ ◯ $\dfrac{2}{5}$

Comparing the two fractions by showing >, <, = symbol:

(1) $\dfrac{6}{9}$ ◯ $\dfrac{3}{9}$ (2) $\dfrac{2}{8}$ ◯ $\dfrac{2}{8}$ (3) $\dfrac{1}{6}$ ◯ $\dfrac{4}{6}$

(4) $\dfrac{3}{6}$ ◯ $\dfrac{2}{6}$ (5) $\dfrac{1}{8}$ ◯ $\dfrac{2}{8}$ (6) $\dfrac{2}{10}$ ◯ $\dfrac{6}{10}$

(7) $\dfrac{1}{9}$ ◯ $\dfrac{3}{9}$ (8) $\dfrac{4}{8}$ ◯ $\dfrac{4}{8}$ (9) $\dfrac{2}{12}$ ◯ $\dfrac{1}{12}$

(10) $\dfrac{2}{12}$ ◯ $\dfrac{5}{12}$ (11) $\dfrac{3}{9}$ ◯ $\dfrac{4}{9}$ (12) $\dfrac{4}{6}$ ◯ $\dfrac{2}{6}$

(13) $\dfrac{1}{2}$ ◯ $\dfrac{1}{2}$ (14) $\dfrac{1}{3}$ ◯ $\dfrac{3}{3}$ (15) $\dfrac{1}{8}$ ◯ $\dfrac{7}{8}$

(16) $\dfrac{7}{8}$ ◯ $\dfrac{3}{8}$ (17) $\dfrac{2}{4}$ ◯ $\dfrac{3}{4}$ (18) $\dfrac{5}{7}$ ◯ $\dfrac{4}{7}$

(19) $\dfrac{4}{6}$ ◯ $\dfrac{1}{6}$ (20) $\dfrac{4}{5}$ ◯ $\dfrac{2}{5}$ (21) $\dfrac{4}{12}$ ◯ $\dfrac{6}{12}$

(22) $\dfrac{5}{7}$ ◯ $\dfrac{4}{7}$ (23) $\dfrac{3}{7}$ ◯ $\dfrac{5}{7}$ (24) $\dfrac{3}{4}$ ◯ $\dfrac{2}{4}$

(25) $\dfrac{2}{5}$ ◯ $\dfrac{4}{5}$ (26) $\dfrac{1}{6}$ ◯ $\dfrac{1}{6}$ (27) $\dfrac{3}{9}$ ◯ $\dfrac{6}{9}$

(28) $\dfrac{1}{4}$ ◯ $\dfrac{3}{4}$ (29) $\dfrac{2}{7}$ ◯ $\dfrac{5}{7}$ (30) $\dfrac{5}{10}$ ◯ $\dfrac{3}{10}$

(31) $\dfrac{3}{9}$ ◯ $\dfrac{6}{9}$ (32) $\dfrac{4}{12}$ ◯ $\dfrac{4}{12}$ (33) $\dfrac{1}{3}$ ◯ $\dfrac{5}{3}$

(34) $\dfrac{2}{7}$ ◯ $\dfrac{4}{7}$ (35) $\dfrac{3}{4}$ ◯ $\dfrac{4}{4}$ (36) $\dfrac{2}{8}$ ◯ $\dfrac{4}{8}$

(37) $\dfrac{1}{3}$ ◯ $\dfrac{2}{3}$ (38) $\dfrac{1}{8}$ ◯ $\dfrac{1}{8}$ (39) $\dfrac{1}{9}$ ◯ $\dfrac{2}{9}$

(40) $\dfrac{5}{10}$ ◯ $\dfrac{4}{10}$ (41) $\dfrac{2}{8}$ ◯ $\dfrac{2}{8}$ (42) $\dfrac{5}{10}$ ◯ $\dfrac{2}{10}$

(43) $\dfrac{3}{5}$ ◯ $\dfrac{1}{5}$ (44) $\dfrac{5}{9}$ ◯ $\dfrac{1}{9}$ (45) $\dfrac{4}{10}$ ◯ $\dfrac{6}{10}$

Comparing the two fractions by showing >, <, = symbol:

(1) $\dfrac{7}{12}$ ○ $\dfrac{5}{12}$ (2) $\dfrac{1}{5}$ ○ $\dfrac{2}{5}$ (3) $\dfrac{1}{2}$ ○ $\dfrac{1}{2}$

(4) $\dfrac{2}{6}$ ○ $\dfrac{3}{6}$ (5) $\dfrac{3}{10}$ ○ $\dfrac{5}{10}$ (6) $\dfrac{4}{6}$ ○ $\dfrac{2}{6}$

(7) $\dfrac{5}{10}$ ○ $\dfrac{4}{10}$ (8) $\dfrac{2}{3}$ ○ $\dfrac{3}{3}$ (9) $\dfrac{2}{8}$ ○ $\dfrac{1}{8}$

(10) $\dfrac{1}{5}$ ○ $\dfrac{1}{5}$ (11) $\dfrac{4}{7}$ ○ $\dfrac{1}{7}$ (12) $\dfrac{3}{4}$ ○ $\dfrac{1}{4}$

(13) $\dfrac{4}{8}$ ○ $\dfrac{2}{8}$ (14) $\dfrac{1}{9}$ ○ $\dfrac{2}{9}$ (15) $\dfrac{3}{10}$ ○ $\dfrac{2}{10}$

(16) $\dfrac{3}{7}$ ○ $\dfrac{3}{7}$ (17) $\dfrac{3}{6}$ ○ $\dfrac{2}{6}$ (18) $\dfrac{5}{9}$ ○ $\dfrac{4}{9}$

(19) $\dfrac{2}{8}$ ○ $\dfrac{2}{8}$ (20) $\dfrac{2}{12}$ ○ $\dfrac{2}{12}$ (21) $\dfrac{3}{7}$ ○ $\dfrac{1}{7}$

(22) $\dfrac{5}{9}$ ○ $\dfrac{5}{9}$ (23) $\dfrac{4}{6}$ ○ $\dfrac{5}{6}$ (24) $\dfrac{4}{12}$ ○ $\dfrac{3}{12}$

(25) $\dfrac{1}{7}$ ○ $\dfrac{5}{7}$ (26) $\dfrac{2}{7}$ ○ $\dfrac{5}{7}$ (27) $\dfrac{5}{8}$ ○ $\dfrac{4}{8}$

(28) $\dfrac{3}{4}$ ○ $\dfrac{3}{4}$ (29) $\dfrac{5}{7}$ ○ $\dfrac{2}{7}$ (30) $\dfrac{2}{5}$ ○ $\dfrac{1}{5}$

(31) $\dfrac{4}{7}$ ○ $\dfrac{5}{7}$ (32) $\dfrac{2}{9}$ ○ $\dfrac{5}{9}$ (33) $\dfrac{5}{6}$ ○ $\dfrac{2}{6}$

(34) $\dfrac{2}{4}$ ○ $\dfrac{2}{4}$ (35) $\dfrac{1}{12}$ ○ $\dfrac{4}{12}$ (36) $\dfrac{1}{3}$ ○ $\dfrac{1}{3}$

(37) $\dfrac{1}{6}$ ○ $\dfrac{2}{6}$ (38) $\dfrac{3}{8}$ ○ $\dfrac{3}{8}$ (39) $\dfrac{5}{7}$ ○ $\dfrac{3}{7}$

(40) $\dfrac{3}{10}$ ○ $\dfrac{4}{10}$ (41) $\dfrac{1}{9}$ ○ $\dfrac{5}{9}$ (42) $\dfrac{3}{9}$ ○ $\dfrac{1}{9}$

(43) $\dfrac{5}{6}$ ○ $\dfrac{3}{6}$ (44) $\dfrac{3}{4}$ ○ $\dfrac{2}{4}$ (45) $\dfrac{2}{7}$ ○ $\dfrac{1}{7}$

Comparing the two fractions by showing >, <, = symbol:

(1) $\dfrac{1}{8}$ ◯ $\dfrac{4}{8}$

(2) $\dfrac{2}{9}$ ◯ $\dfrac{4}{9}$

(3) $\dfrac{4}{8}$ ◯ $\dfrac{3}{8}$

(4) $\dfrac{7}{9}$ ◯ $\dfrac{2}{9}$

(5) $\dfrac{1}{5}$ ◯ $\dfrac{5}{5}$

(6) $\dfrac{3}{12}$ ◯ $\dfrac{2}{12}$

(7) $\dfrac{2}{10}$ ◯ $\dfrac{2}{10}$

(8) $\dfrac{1}{12}$ ◯ $\dfrac{5}{12}$

(9) $\dfrac{2}{6}$ ◯ $\dfrac{1}{6}$

(10) $\dfrac{4}{10}$ ◯ $\dfrac{6}{10}$

(11) $\dfrac{2}{12}$ ◯ $\dfrac{2}{12}$

(12) $\dfrac{4}{10}$ ◯ $\dfrac{2}{10}$

(13) $\dfrac{3}{8}$ ◯ $\dfrac{3}{8}$

(14) $\dfrac{2}{12}$ ◯ $\dfrac{1}{12}$

(15) $\dfrac{1}{8}$ ◯ $\dfrac{1}{8}$

(16) $\dfrac{1}{3}$ ◯ $\dfrac{1}{3}$

(17) $\dfrac{1}{3}$ ◯ $\dfrac{4}{3}$

(18) $\dfrac{5}{12}$ ◯ $\dfrac{4}{12}$

(19) $\dfrac{2}{3}$ ◯ $\dfrac{2}{3}$

(20) $\dfrac{2}{3}$ ◯ $\dfrac{4}{3}$

(21) $\dfrac{4}{9}$ ◯ $\dfrac{1}{9}$

(22) $\dfrac{5}{12}$ ◯ $\dfrac{3}{12}$

(23) $\dfrac{1}{7}$ ◯ $\dfrac{2}{7}$

(24) $\dfrac{2}{4}$ ◯ $\dfrac{1}{4}$

(25) $\dfrac{4}{9}$ ◯ $\dfrac{1}{9}$

(26) $\dfrac{3}{5}$ ◯ $\dfrac{2}{5}$

(27) $\dfrac{3}{6}$ ◯ $\dfrac{2}{6}$

(28) $\dfrac{1}{10}$ ◯ $\dfrac{2}{10}$

(29) $\dfrac{1}{10}$ ◯ $\dfrac{3}{10}$

(30) $\dfrac{4}{7}$ ◯ $\dfrac{3}{7}$

(31) $\dfrac{3}{7}$ ◯ $\dfrac{4}{7}$

(32) $\dfrac{1}{6}$ ◯ $\dfrac{4}{6}$

(33) $\dfrac{1}{10}$ ◯ $\dfrac{1}{10}$

(34) $\dfrac{2}{9}$ ◯ $\dfrac{3}{9}$

(35) $\dfrac{2}{6}$ ◯ $\dfrac{4}{6}$

(36) $\dfrac{3}{8}$ ◯ $\dfrac{2}{8}$

(37) $\dfrac{4}{10}$ ◯ $\dfrac{5}{10}$

(38) $\dfrac{1}{4}$ ◯ $\dfrac{2}{4}$

(39) $\dfrac{4}{5}$ ◯ $\dfrac{2}{5}$

(40) $\dfrac{1}{4}$ ◯ $\dfrac{2}{4}$

(41) $\dfrac{2}{9}$ ◯ $\dfrac{3}{9}$

(42) $\dfrac{1}{7}$ ◯ $\dfrac{1}{7}$

(43) $\dfrac{3}{5}$ ◯ $\dfrac{3}{5}$

(44) $\dfrac{3}{7}$ ◯ $\dfrac{1}{7}$

(45) $\dfrac{2}{9}$ ◯ $\dfrac{1}{9}$

Comparing the two fractions by showing >, <, = symbol:

(1) $\frac{5}{8}$ ◯ $\frac{4}{8}$ (2) $\frac{2}{10}$ ◯ $\frac{4}{10}$ (3) $\frac{4}{12}$ ◯ $\frac{3}{12}$

(4) $\frac{2}{10}$ ◯ $\frac{3}{10}$ (5) $\frac{4}{8}$ ◯ $\frac{1}{8}$ (6) $\frac{5}{12}$ ◯ $\frac{4}{12}$

(7) $\frac{1}{7}$ ◯ $\frac{3}{7}$ (8) $\frac{1}{8}$ ◯ $\frac{3}{8}$ (9) $\frac{3}{12}$ ◯ $\frac{2}{12}$

(10) $\frac{6}{10}$ ◯ $\frac{4}{10}$ (11) $\frac{1}{7}$ ◯ $\frac{6}{7}$ (12) $\frac{2}{12}$ ◯ $\frac{1}{12}$

(13) $\frac{2}{5}$ ◯ $\frac{3}{5}$ (14) $\frac{2}{12}$ ◯ $\frac{3}{12}$ (15) $\frac{1}{9}$ ◯ $\frac{1}{9}$

(16) $\frac{4}{12}$ ◯ $\frac{6}{12}$ (17) $\frac{3}{12}$ ◯ $\frac{1}{12}$ (18) $\frac{3}{5}$ ◯ $\frac{2}{5}$

(19) $\frac{3}{9}$ ◯ $\frac{6}{9}$ (20) $\frac{1}{9}$ ◯ $\frac{3}{9}$ (21) $\frac{2}{10}$ ◯ $\frac{1}{10}$

(22) $\frac{4}{10}$ ◯ $\frac{6}{10}$ (23) $\frac{2}{7}$ ◯ $\frac{5}{7}$ (24) $\frac{4}{4}$ ◯ $\frac{3}{4}$

(25) $\frac{1}{4}$ ◯ $\frac{4}{4}$ (26) $\frac{1}{12}$ ◯ $\frac{3}{12}$ (27) $\frac{1}{12}$ ◯ $\frac{1}{12}$

(28) $\frac{2}{7}$ ◯ $\frac{2}{7}$ (29) $\frac{1}{5}$ ◯ $\frac{3}{5}$ (30) $\frac{3}{3}$ ◯ $\frac{2}{3}$

(31) $\frac{1}{2}$ ◯ $\frac{3}{2}$ (32) $\frac{2}{8}$ ◯ $\frac{5}{8}$ (33) $\frac{4}{8}$ ◯ $\frac{2}{8}$

(34) $\frac{2}{10}$ ◯ $\frac{5}{10}$ (35) $\frac{1}{6}$ ◯ $\frac{5}{6}$ (36) $\frac{5}{10}$ ◯ $\frac{4}{10}$

(37) $\frac{1}{8}$ ◯ $\frac{1}{8}$ (38) $\frac{2}{12}$ ◯ $\frac{5}{12}$ (39) $\frac{2}{12}$ ◯ $\frac{1}{12}$

(40) $\frac{3}{12}$ ◯ $\frac{3}{12}$ (41) $\frac{4}{10}$ ◯ $\frac{2}{10}$ (42) $\frac{1}{6}$ ◯ $\frac{1}{6}$

(43) $\frac{2}{3}$ ◯ $\frac{2}{3}$ (44) $\frac{3}{9}$ ◯ $\frac{6}{9}$ (45) $\frac{3}{7}$ ◯ $\frac{2}{7}$

Comparing the two fractions by showing >, <, = symbol:

(1) $\frac{3}{8}$ ◯ $\frac{2}{8}$ (2) $\frac{4}{7}$ ◯ $\frac{1}{7}$ (3) $\frac{4}{6}$ ◯ $\frac{3}{6}$

(4) $\frac{5}{6}$ ◯ $\frac{1}{6}$ (5) $\frac{3}{8}$ ◯ $\frac{5}{8}$ (6) $\frac{2}{3}$ ◯ $\frac{1}{3}$

(7) $\frac{1}{10}$ ◯ $\frac{5}{10}$ (8) $\frac{2}{6}$ ◯ $\frac{4}{6}$ (9) $\frac{1}{12}$ ◯ $\frac{1}{12}$

(10) $\frac{7}{9}$ ◯ $\frac{5}{9}$ (11) $\frac{1}{10}$ ◯ $\frac{4}{10}$ (12) $\frac{3}{4}$ ◯ $\frac{2}{4}$

(13) $\frac{3}{5}$ ◯ $\frac{2}{5}$ (14) $\frac{2}{5}$ ◯ $\frac{3}{5}$ (15) $\frac{5}{8}$ ◯ $\frac{5}{8}$

(16) $\frac{2}{6}$ ◯ $\frac{5}{6}$ (17) $\frac{4}{9}$ ◯ $\frac{1}{9}$ (18) $\frac{1}{7}$ ◯ $\frac{3}{7}$

(19) $\frac{1}{3}$ ◯ $\frac{2}{3}$ (20) $\frac{1}{12}$ ◯ $\frac{5}{12}$ (21) $\frac{3}{8}$ ◯ $\frac{5}{8}$

(22) $\frac{4}{7}$ ◯ $\frac{4}{7}$ (23) $\frac{3}{12}$ ◯ $\frac{2}{12}$ (24) $\frac{2}{12}$ ◯ $\frac{3}{12}$

(25) $\frac{3}{4}$ ◯ $\frac{1}{4}$ (26) $\frac{2}{9}$ ◯ $\frac{5}{9}$ (27) $\frac{1}{6}$ ◯ $\frac{2}{6}$

(28) $\frac{5}{8}$ ◯ $\frac{2}{8}$ (29) $\frac{1}{8}$ ◯ $\frac{2}{8}$ (30) $\frac{2}{9}$ ◯ $\frac{7}{9}$

(31) $\frac{2}{4}$ ◯ $\frac{4}{4}$ (32) $\frac{3}{6}$ ◯ $\frac{2}{6}$ (33) $\frac{1}{12}$ ◯ $\frac{1}{12}$

(34) $\frac{1}{9}$ ◯ $\frac{4}{9}$ (35) $\frac{4}{12}$ ◯ $\frac{3}{12}$ (36) $\frac{1}{8}$ ◯ $\frac{7}{8}$

(37) $\frac{4}{10}$ ◯ $\frac{2}{10}$ (38) $\frac{2}{7}$ ◯ $\frac{4}{7}$ (39) $\frac{3}{6}$ ◯ $\frac{2}{6}$

(40) $\frac{3}{10}$ ◯ $\frac{4}{10}$ (41) $\frac{3}{10}$ ◯ $\frac{1}{10}$ (42) $\frac{2}{3}$ ◯ $\frac{3}{3}$

(43) $\frac{1}{6}$ ◯ $\frac{3}{6}$ (44) $\frac{4}{6}$ ◯ $\frac{1}{6}$ (45) $\frac{1}{4}$ ◯ $\frac{3}{4}$

Comparing the two fractions by showing >, <, = symbol:

(1) $\dfrac{2}{8}$ ◯ $\dfrac{4}{8}$ (2) $\dfrac{2}{8}$ ◯ $\dfrac{4}{8}$ (3) $\dfrac{1}{9}$ ◯ $\dfrac{8}{9}$

(4) $\dfrac{3}{6}$ ◯ $\dfrac{3}{6}$ (5) $\dfrac{1}{7}$ ◯ $\dfrac{4}{7}$ (6) $\dfrac{4}{10}$ ◯ $\dfrac{1}{10}$

(7) $\dfrac{1}{7}$ ◯ $\dfrac{1}{7}$ (8) $\dfrac{3}{12}$ ◯ $\dfrac{2}{12}$ (9) $\dfrac{1}{2}$ ◯ $\dfrac{4}{2}$

(10) $\dfrac{2}{5}$ ◯ $\dfrac{2}{5}$ (11) $\dfrac{1}{12}$ ◯ $\dfrac{4}{12}$ (12) $\dfrac{4}{8}$ ◯ $\dfrac{4}{8}$

(13) $\dfrac{4}{9}$ ◯ $\dfrac{2}{9}$ (14) $\dfrac{1}{6}$ ◯ $\dfrac{3}{6}$ (15) $\dfrac{1}{5}$ ◯ $\dfrac{4}{5}$

(16) $\dfrac{1}{5}$ ◯ $\dfrac{4}{5}$ (17) $\dfrac{2}{10}$ ◯ $\dfrac{3}{10}$ (18) $\dfrac{1}{10}$ ◯ $\dfrac{4}{10}$

(19) $\dfrac{5}{7}$ ◯ $\dfrac{2}{7}$ (20) $\dfrac{3}{9}$ ◯ $\dfrac{1}{9}$ (21) $\dfrac{2}{10}$ ◯ $\dfrac{3}{10}$

(22) $\dfrac{3}{10}$ ◯ $\dfrac{3}{10}$ (23) $\dfrac{1}{5}$ ◯ $\dfrac{4}{5}$ (24) $\dfrac{1}{12}$ ◯ $\dfrac{2}{12}$

(25) $\dfrac{2}{12}$ ◯ $\dfrac{4}{12}$ (26) $\dfrac{4}{12}$ ◯ $\dfrac{1}{12}$ (27) $\dfrac{1}{8}$ ◯ $\dfrac{5}{8}$

(28) $\dfrac{1}{10}$ ◯ $\dfrac{3}{10}$ (29) $\dfrac{2}{12}$ ◯ $\dfrac{3}{12}$ (30) $\dfrac{3}{5}$ ◯ $\dfrac{1}{5}$

(31) $\dfrac{4}{6}$ ◯ $\dfrac{2}{6}$ (32) $\dfrac{3}{7}$ ◯ $\dfrac{4}{7}$ (33) $\dfrac{3}{9}$ ◯ $\dfrac{5}{9}$

(34) $\dfrac{3}{7}$ ◯ $\dfrac{1}{7}$ (35) $\dfrac{1}{9}$ ◯ $\dfrac{4}{9}$ (36) $\dfrac{1}{6}$ ◯ $\dfrac{5}{6}$

(37) $\dfrac{2}{9}$ ◯ $\dfrac{2}{9}$ (38) $\dfrac{1}{8}$ ◯ $\dfrac{5}{8}$ (39) $\dfrac{2}{6}$ ◯ $\dfrac{5}{6}$

(40) $\dfrac{1}{12}$ ◯ $\dfrac{3}{12}$ (41) $\dfrac{2}{6}$ ◯ $\dfrac{5}{6}$ (42) $\dfrac{1}{12}$ ◯ $\dfrac{5}{12}$

(43) $\dfrac{5}{10}$ ◯ $\dfrac{3}{10}$ (44) $\dfrac{3}{12}$ ◯ $\dfrac{1}{12}$ (45) $\dfrac{1}{4}$ ◯ $\dfrac{5}{4}$

Comparing the two fractions by showing >, <, = symbol:

(1) $\dfrac{2}{10}$ ◯ $\dfrac{2}{10}$ (2) $\dfrac{4}{8}$ ◯ $\dfrac{3}{8}$ (3) $\dfrac{1}{7}$ ◯ $\dfrac{4}{7}$

(4) $\dfrac{1}{8}$ ◯ $\dfrac{2}{8}$ (5) $\dfrac{1}{10}$ ◯ $\dfrac{3}{10}$ (6) $\dfrac{4}{6}$ ◯ $\dfrac{2}{6}$

(7) $\dfrac{4}{8}$ ◯ $\dfrac{4}{8}$ (8) $\dfrac{2}{7}$ ◯ $\dfrac{3}{7}$ (9) $\dfrac{3}{12}$ ◯ $\dfrac{3}{12}$

(10) $\dfrac{3}{9}$ ◯ $\dfrac{4}{9}$ (11) $\dfrac{1}{12}$ ◯ $\dfrac{4}{12}$ (12) $\dfrac{1}{3}$ ◯ $\dfrac{3}{3}$

(13) $\dfrac{1}{3}$ ◯ $\dfrac{3}{3}$ (14) $\dfrac{1}{5}$ ◯ $\dfrac{2}{5}$ (15) $\dfrac{1}{5}$ ◯ $\dfrac{3}{5}$

(16) $\dfrac{2}{4}$ ◯ $\dfrac{3}{4}$ (17) $\dfrac{4}{9}$ ◯ $\dfrac{3}{9}$ (18) $\dfrac{2}{5}$ ◯ $\dfrac{5}{5}$

(19) $\dfrac{4}{5}$ ◯ $\dfrac{2}{5}$ (20) $\dfrac{3}{8}$ ◯ $\dfrac{4}{8}$ (21) $\dfrac{1}{10}$ ◯ $\dfrac{5}{10}$

(22) $\dfrac{3}{7}$ ◯ $\dfrac{5}{7}$ (23) $\dfrac{2}{12}$ ◯ $\dfrac{4}{12}$ (24) $\dfrac{1}{9}$ ◯ $\dfrac{2}{9}$

(25) $\dfrac{1}{6}$ ◯ $\dfrac{1}{6}$ (26) $\dfrac{4}{7}$ ◯ $\dfrac{3}{7}$ (27) $\dfrac{2}{7}$ ◯ $\dfrac{2}{7}$

(28) $\dfrac{2}{7}$ ◯ $\dfrac{5}{7}$ (29) $\dfrac{1}{9}$ ◯ $\dfrac{2}{9}$ (30) $\dfrac{1}{12}$ ◯ $\dfrac{1}{12}$

(31) $\dfrac{4}{12}$ ◯ $\dfrac{4}{12}$ (32) $\dfrac{2}{5}$ ◯ $\dfrac{4}{5}$ (33) $\dfrac{1}{7}$ ◯ $\dfrac{5}{7}$

(34) $\dfrac{3}{4}$ ◯ $\dfrac{4}{4}$ (35) $\dfrac{1}{3}$ ◯ $\dfrac{2}{3}$ (36) $\dfrac{2}{8}$ ◯ $\dfrac{2}{8}$

(37) $\dfrac{1}{10}$ ◯ $\dfrac{1}{10}$ (38) $\dfrac{2}{5}$ ◯ $\dfrac{1}{5}$ (39) $\dfrac{1}{3}$ ◯ $\dfrac{5}{3}$

(40) $\dfrac{2}{8}$ ◯ $\dfrac{2}{8}$ (41) $\dfrac{1}{6}$ ◯ $\dfrac{5}{6}$ (42) $\dfrac{2}{3}$ ◯ $\dfrac{5}{3}$

(43) $\dfrac{5}{9}$ ◯ $\dfrac{1}{9}$ (44) $\dfrac{3}{4}$ ◯ $\dfrac{1}{4}$ (45) $\dfrac{1}{6}$ ◯ $\dfrac{3}{6}$

Reduce the fractions to the simplest form :

(1) $\dfrac{3}{6} = \underline{\quad}$

(2) $\dfrac{50}{100} = \underline{\quad}$

(3) $\dfrac{30}{40} = \underline{\quad}$

(4) $\dfrac{8}{12} = \underline{\quad}$

(5) $\dfrac{8}{40} = \underline{\quad}$

(6) $\dfrac{63}{81} = \underline{\quad}$

(7) $\dfrac{2}{7} = \underline{\quad}$

(8) $\dfrac{56}{70} = \underline{\quad}$

(9) $\dfrac{84}{96} = \underline{\quad}$

(10) $\dfrac{9}{12} = \underline{\quad}$

(11) $\dfrac{10}{30} = \underline{\quad}$

(12) $\dfrac{48}{72} = \underline{\quad}$

(13) $\dfrac{14}{28} = \underline{\quad}$

(14) $\dfrac{24}{48} = \underline{\quad}$

(15) $\dfrac{30}{90} = \underline{\quad}$

(16) $\dfrac{25}{100} = \underline{\quad}$

(17) $\dfrac{42}{56} = \underline{\quad}$

(18) $\dfrac{35}{105} = \underline{\quad}$

(19) $\dfrac{12}{30} = \underline{\quad}$

(20) $\dfrac{72}{96} = \underline{\quad}$

(21) $\dfrac{70}{100} = \underline{\quad}$

(22) $\dfrac{16}{48} = \underline{\quad}$

(23) $\dfrac{7}{28} = \underline{\quad}$

(24) $\dfrac{44}{55} = \underline{\quad}$

(25) $\dfrac{20}{50} = \underline{\quad}$

(26) $\dfrac{16}{80} = \underline{\quad}$

(27) $\dfrac{60}{75} = \underline{\quad}$

(28) $\dfrac{36}{48} = \underline{\quad}$

(29) $\dfrac{48}{72} = \underline{\quad}$

(30) $\dfrac{28}{56} = \underline{\quad}$

(31) $\dfrac{21}{28} = \underline{\quad}$

(32) $\dfrac{50}{60} = \underline{\quad}$

(33) $\dfrac{21}{28} = \underline{\quad}$

(34) $\dfrac{42}{63} = \underline{\quad}$

(35) $\dfrac{64}{80} = \underline{\quad}$

(36) $\dfrac{25}{100} = \underline{\quad}$

(37) $\dfrac{15}{75} = \underline{\quad}$

(38) $\dfrac{25}{50} = \underline{\quad}$

(39) $\dfrac{24}{30} = \underline{\quad}$

(40) $\dfrac{24}{60} = \underline{\quad}$

(41) $\dfrac{21}{63} = \underline{\quad}$

(42) $\dfrac{27}{63} = \underline{\quad}$

(43) $\dfrac{18}{36} = \underline{\quad}$

(44) $\dfrac{12}{60} = \underline{\quad}$

(45) $\dfrac{32}{80} = \underline{\quad}$

Reduce the fractions to the simplest form :

(1) $\dfrac{28}{35} = \dfrac{\quad}{\quad}$ (2) $\dfrac{16}{64} = \dfrac{\quad}{\quad}$ (3) $\dfrac{48}{80} = \dfrac{\quad}{\quad}$

(4) $\dfrac{35}{70} = \dfrac{\quad}{\quad}$ (5) $\dfrac{32}{96} = \dfrac{\quad}{\quad}$ (6) $\dfrac{18}{90} = \dfrac{\quad}{\quad}$

(7) $\dfrac{45}{90} = \dfrac{\quad}{\quad}$ (8) $\dfrac{30}{100} = \dfrac{\quad}{\quad}$ (9) $\dfrac{32}{48} = \dfrac{\quad}{\quad}$

(10) $\dfrac{56}{84} = \dfrac{\quad}{\quad}$ (11) $\dfrac{12}{84} = \dfrac{\quad}{\quad}$ (12) $\dfrac{42}{84} = \dfrac{\quad}{\quad}$

(13) $\dfrac{72}{96} = \dfrac{\quad}{\quad}$ (14) $\dfrac{18}{54} = \dfrac{\quad}{\quad}$ (15) $\dfrac{54}{99} = \dfrac{\quad}{\quad}$

(16) $\dfrac{27}{36} = \dfrac{\quad}{\quad}$ (17) $\dfrac{27}{81} = \dfrac{\quad}{\quad}$ (18) $\dfrac{12}{40} = \dfrac{\quad}{\quad}$

(19) $\dfrac{22}{44} = \dfrac{\quad}{\quad}$ (20) $\dfrac{48}{72} = \dfrac{\quad}{\quad}$ (21) $\dfrac{72}{84} = \dfrac{\quad}{\quad}$

(22) $\dfrac{32}{64} = \dfrac{\quad}{\quad}$ (23) $\dfrac{10}{90} = \dfrac{\quad}{\quad}$ (24) $\dfrac{70}{80} = \dfrac{\quad}{\quad}$

(25) $\dfrac{24}{36} = \dfrac{\quad}{\quad}$ (26) $\dfrac{40}{80} = \dfrac{\quad}{\quad}$ (27) $\dfrac{80}{90} = \dfrac{\quad}{\quad}$

(28) $\dfrac{63}{84} = \dfrac{\quad}{\quad}$ (29) $\dfrac{80}{100} = \dfrac{\quad}{\quad}$ (30) $\dfrac{63}{90} = \dfrac{\quad}{\quad}$

(31) $\dfrac{20}{25} = \dfrac{\quad}{\quad}$ (32) $\dfrac{72}{96} = \dfrac{\quad}{\quad}$ (33) $\dfrac{32}{64} = \dfrac{\quad}{\quad}$

(34) $\dfrac{22}{77} = \dfrac{\quad}{\quad}$ (35) $\dfrac{42}{84} = \dfrac{\quad}{\quad}$ (36) $\dfrac{84}{98} = \dfrac{\quad}{\quad}$

(37) $\dfrac{40}{60} = \dfrac{\quad}{\quad}$ (38) $\dfrac{20}{80} = \dfrac{\quad}{\quad}$ (39) $\dfrac{18}{24} = \dfrac{\quad}{\quad}$

(40) $\dfrac{9}{45} = \dfrac{\quad}{\quad}$ (41) $\dfrac{63}{84} = \dfrac{\quad}{\quad}$ (42) $\dfrac{40}{60} = \dfrac{\quad}{\quad}$

(43) $\dfrac{54}{90} = \dfrac{\quad}{\quad}$ (44) $\dfrac{7}{77} = \dfrac{\quad}{\quad}$ (45) $\dfrac{36}{45} = \dfrac{\quad}{\quad}$

Reduce the fractions to the simplest form :

(1) $\dfrac{77}{91} = \underline{\quad}$

(2) $\dfrac{9}{54} = \underline{\quad}$

(3) $\dfrac{80}{100} = \underline{\quad}$

(4) $\dfrac{68}{85} = \underline{\quad}$

(5) $\dfrac{32}{80} = \underline{\quad}$

(6) $\dfrac{15}{75} = \underline{\quad}$

(7) $\dfrac{80}{100} = \underline{\quad}$

(8) $\dfrac{36}{72} = \underline{\quad}$

(9) $\dfrac{42}{56} = \underline{\quad}$

(10) $\dfrac{16}{32} = \underline{\quad}$

(11) $\dfrac{25}{75} = \underline{\quad}$

(12) $\dfrac{15}{45} = \underline{\quad}$

(13) $\dfrac{25}{75} = \underline{\quad}$

(14) $\dfrac{15}{30} = \underline{\quad}$

(15) $\dfrac{63}{81} = \underline{\quad}$

(16) $\dfrac{36}{48} = \underline{\quad}$

(17) $\dfrac{24}{72} = \underline{\quad}$

(18) $\dfrac{35}{70} = \underline{\quad}$

(19) $\dfrac{28}{70} = \underline{\quad}$

(20) $\dfrac{40}{100} = \underline{\quad}$

(21) $\dfrac{12}{60} = \underline{\quad}$

(22) $\dfrac{18}{24} = \underline{\quad}$

(23) $\dfrac{21}{42} = \underline{\quad}$

(24) $\dfrac{28}{35} = \underline{\quad}$

(25) $\dfrac{64}{80} = \underline{\quad}$

(26) $\dfrac{10}{20} = \underline{\quad}$

(27) $\dfrac{56}{80} = \underline{\quad}$

(28) $\dfrac{63}{126} = \underline{\quad}$

(29) $\dfrac{35}{70} = \underline{\quad}$

(30) $\dfrac{8}{80} = \underline{\quad}$

(31) $\dfrac{56}{84} = \underline{\quad}$

(32) $\dfrac{6}{54} = \underline{\quad}$

(33) $\dfrac{16}{80} = \underline{\quad}$

(34) $\dfrac{77}{154} = \underline{\quad}$

(35) $\dfrac{6}{42} = \underline{\quad}$

(36) $\dfrac{45}{60} = \underline{\quad}$

(37) $\dfrac{48}{60} = \underline{\quad}$

(38) $\dfrac{16}{48} = \underline{\quad}$

(39) $\dfrac{48}{60} = \underline{\quad}$

(40) $\dfrac{45}{75} = \underline{\quad}$

(41) $\dfrac{56}{84} = \underline{\quad}$

(42) $\dfrac{32}{40} = \underline{\quad}$

(43) $\dfrac{32}{64} = \underline{\quad}$

(44) $\dfrac{45}{90} = \underline{\quad}$

(45) $\dfrac{48}{72} = \underline{\quad}$

Reduce the fractions to the simplest form :

(1) $\dfrac{54}{72} = \underline{\quad}$

(2) $\dfrac{8}{64} = \underline{\quad}$

(3) $\dfrac{42}{70} = \underline{\quad}$

(4) $\dfrac{25}{50} = \underline{\quad}$

(5) $\dfrac{63}{90} = \underline{\quad}$

(6) $\dfrac{16}{48} = \underline{\quad}$

(7) $\dfrac{60}{80} = \underline{\quad}$

(8) $\dfrac{36}{48} = \underline{\quad}$

(9) $\dfrac{35}{105} = \underline{\quad}$

(10) $\dfrac{84}{98} = \underline{\quad}$

(11) $\dfrac{24}{48} = \underline{\quad}$

(12) $\dfrac{60}{90} = \underline{\quad}$

(13) $\dfrac{63}{105} = \underline{\quad}$

(14) $\dfrac{56}{80} = \underline{\quad}$

(15) $\dfrac{18}{36} = \underline{\quad}$

(16) $\dfrac{12}{30} = \underline{\quad}$

(17) $\dfrac{18}{90} = \underline{\quad}$

(18) $\dfrac{10}{50} = \underline{\quad}$

(19) $\dfrac{99}{132} = \underline{\quad}$

(20) $\dfrac{44}{88} = \underline{\quad}$

(21) $\dfrac{56}{84} = \underline{\quad}$

(22) $\dfrac{42}{63} = \underline{\quad}$

(23) $\dfrac{14}{28} = \underline{\quad}$

(24) $\dfrac{20}{80} = \underline{\quad}$

(25) $\dfrac{44}{88} = \underline{\quad}$

(26) $\dfrac{20}{60} = \underline{\quad}$

(27) $\dfrac{50}{75} = \underline{\quad}$

(28) $\dfrac{72}{90} = \underline{\quad}$

(29) $\dfrac{49}{98} = \underline{\quad}$

(30) $\dfrac{54}{72} = \underline{\quad}$

(31) $\dfrac{42}{84} = \underline{\quad}$

(32) $\dfrac{18}{36} = \underline{\quad}$

(33) $\dfrac{72}{84} = \underline{\quad}$

(34) $\dfrac{60}{96} = \underline{\quad}$

(35) $\dfrac{75}{100} = \underline{\quad}$

(36) $\dfrac{45}{75} = \underline{\quad}$

(37) $\dfrac{20}{80} = \underline{\quad}$

(38) $\dfrac{42}{70} = \underline{\quad}$

(39) $\dfrac{25}{75} = \underline{\quad}$

(40) $\dfrac{30}{90} = \underline{\quad}$

(41) $\dfrac{45}{60} = \underline{\quad}$

(42) $\dfrac{50}{70} = \underline{\quad}$

(43) $\dfrac{81}{108} = \underline{\quad}$

(44) $\dfrac{64}{96} = \underline{\quad}$

(45) $\dfrac{66}{99} = \underline{\quad}$

Reduce the fractions to the simplest form :

(1) $\dfrac{77}{110} = \underline{\quad}$

(2) $\dfrac{56}{84} = \underline{\quad}$

(3) $\dfrac{40}{80} = \underline{\quad}$

(4) $\dfrac{84}{105} = \underline{\quad}$

(5) $\dfrac{36}{72} = \underline{\quad}$

(6) $\dfrac{42}{49} = \underline{\quad}$

(7) $\dfrac{28}{35} = \underline{\quad}$

(8) $\dfrac{48}{96} = \underline{\quad}$

(9) $\dfrac{30}{90} = \underline{\quad}$

(10) $\dfrac{80}{120} = \underline{\quad}$

(11) $\dfrac{27}{81} = \underline{\quad}$

(12) $\dfrac{64}{80} = \underline{\quad}$

(13) $\dfrac{39}{52} = \underline{\quad}$

(14) $\dfrac{20}{25} = \underline{\quad}$

(15) $\dfrac{18}{72} = \underline{\quad}$

(16) $\dfrac{42}{70} = \underline{\quad}$

(17) $\dfrac{24}{60} = \underline{\quad}$

(18) $\dfrac{45}{60} = \underline{\quad}$

(19) $\dfrac{54}{81} = \underline{\quad}$

(20) $\dfrac{32}{64} = \underline{\quad}$

(21) $\dfrac{24}{36} = \underline{\quad}$

(22) $\dfrac{66}{99} = \underline{\quad}$

(23) $\dfrac{45}{75} = \underline{\quad}$

(24) $\dfrac{35}{50} = \underline{\quad}$

(25) $\dfrac{48}{64} = \underline{\quad}$

(26) $\dfrac{36}{54} = \underline{\quad}$

(27) $\dfrac{27}{45} = \underline{\quad}$

(28) $\dfrac{90}{100} = \underline{\quad}$

(29) $\dfrac{6}{66} = \underline{\quad}$

(30) $\dfrac{28}{42} = \underline{\quad}$

(31) $\dfrac{48}{72} = \underline{\quad}$

(32) $\dfrac{24}{72} = \underline{\quad}$

(33) $\dfrac{10}{20} = \underline{\quad}$

(34) $\dfrac{15}{30} = \underline{\quad}$

(35) $\dfrac{50}{75} = \underline{\quad}$

(36) $\dfrac{56}{70} = \underline{\quad}$

(37) $\dfrac{7}{56} = \underline{\quad}$

(38) $\dfrac{21}{42} = \underline{\quad}$

(39) $\dfrac{20}{25} = \underline{\quad}$

(40) $\dfrac{16}{80} = \underline{\quad}$

(41) $\dfrac{14}{42} = \underline{\quad}$

(42) $\dfrac{56}{98} = \underline{\quad}$

(43) $\dfrac{54}{90} = \underline{\quad}$

(44) $\dfrac{32}{80} = \underline{\quad}$

(45) $\dfrac{12}{48} = \underline{\quad}$

Reduce the fractions to the simplest form :

(1) $\dfrac{72}{80} = \underline{\quad}$ (2) $\dfrac{50}{100} = \underline{\quad}$ (3) $\dfrac{16}{80} = \underline{\quad}$

(4) $\dfrac{21}{70} = \underline{\quad}$ (5) $\dfrac{35}{70} = \underline{\quad}$ (6) $\dfrac{25}{50} = \underline{\quad}$

(7) $\dfrac{35}{70} = \underline{\quad}$ (8) $\dfrac{9}{45} = \underline{\quad}$ (9) $\dfrac{8}{72} = \underline{\quad}$

(10) $\dfrac{27}{45} = \underline{\quad}$ (11) $\dfrac{22}{88} = \underline{\quad}$ (12) $\dfrac{40}{90} = \underline{\quad}$

(13) $\dfrac{22}{88} = \underline{\quad}$ (14) $\dfrac{80}{100} = \underline{\quad}$ (15) $\dfrac{21}{42} = \underline{\quad}$

(16) $\dfrac{63}{90} = \underline{\quad}$ (17) $\dfrac{30}{60} = \underline{\quad}$ (18) $\dfrac{30}{60} = \underline{\quad}$

(19) $\dfrac{38}{76} = \underline{\quad}$ (20) $\dfrac{16}{32} = \underline{\quad}$ (21) $\dfrac{12}{24} = \underline{\quad}$

(22) $\dfrac{66}{88} = \underline{\quad}$ (23) $\dfrac{25}{100} = \underline{\quad}$ (24) $\dfrac{15}{30} = \underline{\quad}$

(25) $\dfrac{26}{78} = \underline{\quad}$ (26) $\dfrac{42}{84} = \underline{\quad}$ (27) $\dfrac{16}{64} = \underline{\quad}$

(28) $\dfrac{18}{54} = \underline{\quad}$ (29) $\dfrac{28}{84} = \underline{\quad}$ (30) $\dfrac{36}{72} = \underline{\quad}$

(31) $\dfrac{72}{96} = \underline{\quad}$ (32) $\dfrac{45}{90} = \underline{\quad}$ (33) $\dfrac{44}{88} = \underline{\quad}$

(34) $\dfrac{33}{44} = \underline{\quad}$ (35) $\dfrac{39}{78} = \underline{\quad}$ (36) $\dfrac{75}{100} = \underline{\quad}$

(37) $\dfrac{40}{80} = \underline{\quad}$ (38) $\dfrac{30}{50} = \underline{\quad}$ (39) $\dfrac{32}{96} = \underline{\quad}$

(40) $\dfrac{27}{54} = \underline{\quad}$ (41) $\dfrac{28}{70} = \underline{\quad}$ (42) $\dfrac{54}{90} = \underline{\quad}$

(43) $\dfrac{14}{35} = \underline{\quad}$ (44) $\dfrac{80}{96} = \underline{\quad}$ (45) $\dfrac{72}{96} = \underline{\quad}$

Reduce the fractions to the simplest form :

(1) $\dfrac{56}{70} = \underline{\quad}$

(2) $\dfrac{36}{45} = \underline{\quad}$

(3) $\dfrac{63}{84} = \underline{\quad}$

(4) $\dfrac{25}{50} = \underline{\quad}$

(5) $\dfrac{15}{75} = \underline{\quad}$

(6) $\dfrac{32}{40} = \underline{\quad}$

(7) $\dfrac{80}{100} = \underline{\quad}$

(8) $\dfrac{24}{48} = \underline{\quad}$

(9) $\dfrac{28}{35} = \underline{\quad}$

(10) $\dfrac{55}{77} = \underline{\quad}$

(11) $\dfrac{10}{90} = \underline{\quad}$

(12) $\dfrac{12}{15} = \underline{\quad}$

(13) $\dfrac{51}{68} = \underline{\quad}$

(14) $\dfrac{6}{30} = \underline{\quad}$

(15) $\dfrac{60}{70} = \underline{\quad}$

(16) $\dfrac{48}{60} = \underline{\quad}$

(17) $\dfrac{6}{60} = \underline{\quad}$

(18) $\dfrac{80}{90} = \underline{\quad}$

(19) $\dfrac{63}{84} = \underline{\quad}$

(20) $\dfrac{28}{56} = \underline{\quad}$

(21) $\dfrac{40}{100} = \underline{\quad}$

(22) $\dfrac{7}{35} = \underline{\quad}$

(23) $\dfrac{16}{64} = \underline{\quad}$

(24) $\dfrac{27}{54} = \underline{\quad}$

(25) $\dfrac{36}{90} = \underline{\quad}$

(26) $\dfrac{63}{90} = \underline{\quad}$

(27) $\dfrac{56}{72} = \underline{\quad}$

(28) $\dfrac{6}{54} = \underline{\quad}$

(29) $\dfrac{48}{80} = \underline{\quad}$

(30) $\dfrac{15}{45} = \underline{\quad}$

(31) $\dfrac{10}{20} = \underline{\quad}$

(32) $\dfrac{45}{60} = \underline{\quad}$

(33) $\dfrac{77}{88} = \underline{\quad}$

(34) $\dfrac{72}{84} = \underline{\quad}$

(35) $\dfrac{70}{100} = \underline{\quad}$

(36) $\dfrac{21}{84} = \underline{\quad}$

(37) $\dfrac{27}{81} = \underline{\quad}$

(38) $\dfrac{12}{84} = \underline{\quad}$

(39) $\dfrac{18}{90} = \underline{\quad}$

(40) $\dfrac{48}{80} = \underline{\quad}$

(41) $\dfrac{35}{105} = \underline{\quad}$

(42) $\dfrac{40}{100} = \underline{\quad}$

(43) $\dfrac{63}{72} = \underline{\quad}$

(44) $\dfrac{30}{90} = \underline{\quad}$

(45) $\dfrac{35}{70} = \underline{\quad}$

Reduce the fractions to the simplest form :

(1) $\dfrac{64}{80} = \underline{\hphantom{--}}$

(2) $\dfrac{18}{90} = \underline{\hphantom{--}}$

(3) $\dfrac{80}{90} = \underline{\hphantom{--}}$

(4) $\dfrac{8}{64} = \underline{\hphantom{--}}$

(5) $\dfrac{42}{63} = \underline{\hphantom{--}}$

(6) $\dfrac{32}{64} = \underline{\hphantom{--}}$

(7) $\dfrac{25}{75} = \underline{\hphantom{--}}$

(8) $\dfrac{16}{80} = \underline{\hphantom{--}}$

(9) $\dfrac{30}{50} = \underline{\hphantom{--}}$

(10) $\dfrac{18}{72} = \underline{\hphantom{--}}$

(11) $\dfrac{36}{54} = \underline{\hphantom{--}}$

(12) $\dfrac{44}{88} = \underline{\hphantom{--}}$

(13) $\dfrac{15}{60} = \underline{\hphantom{--}}$

(14) $\dfrac{15}{45} = \underline{\hphantom{--}}$

(15) $\dfrac{24}{48} = \underline{\hphantom{--}}$

(16) $\dfrac{40}{100} = \underline{\hphantom{--}}$

(17) $\dfrac{16}{48} = \underline{\hphantom{--}}$

(18) $\dfrac{15}{75} = \underline{\hphantom{--}}$

(19) $\dfrac{56}{84} = \underline{\hphantom{--}}$

(20) $\dfrac{42}{56} = \underline{\hphantom{--}}$

(21) $\dfrac{54}{60} = \underline{\hphantom{--}}$

(22) $\dfrac{12}{84} = \underline{\hphantom{--}}$

(23) $\dfrac{72}{96} = \underline{\hphantom{--}}$

(24) $\dfrac{56}{84} = \underline{\hphantom{--}}$

(25) $\dfrac{24}{96} = \underline{\hphantom{--}}$

(26) $\dfrac{63}{84} = \underline{\hphantom{--}}$

(27) $\dfrac{42}{84} = \underline{\hphantom{--}}$

(28) $\dfrac{32}{64} = \underline{\hphantom{--}}$

(29) $\dfrac{18}{72} = \underline{\hphantom{--}}$

(30) $\dfrac{18}{72} = \underline{\hphantom{--}}$

(31) $\dfrac{10}{40} = \underline{\hphantom{--}}$

(32) $\dfrac{56}{70} = \underline{\hphantom{--}}$

(33) $\dfrac{77}{91} = \underline{\hphantom{--}}$

(34) $\dfrac{28}{84} = \underline{\hphantom{--}}$

(35) $\dfrac{64}{80} = \underline{\hphantom{--}}$

(36) $\dfrac{63}{84} = \underline{\hphantom{--}}$

(37) $\dfrac{15}{45} = \underline{\hphantom{--}}$

(38) $\dfrac{21}{28} = \underline{\hphantom{--}}$

(39) $\dfrac{32}{48} = \underline{\hphantom{--}}$

(40) $\dfrac{18}{45} = \underline{\hphantom{--}}$

(41) $\dfrac{77}{88} = \underline{\hphantom{--}}$

(42) $\dfrac{16}{64} = \underline{\hphantom{--}}$

(43) $\dfrac{14}{70} = \underline{\hphantom{--}}$

(44) $\dfrac{80}{90} = \underline{\hphantom{--}}$

(45) $\dfrac{28}{84} = \underline{\hphantom{--}}$

Reduce the fractions to the simplest form :

(1) $\dfrac{33}{99} = \text{—}$ (2) $\dfrac{54}{72} = \text{—}$ (3) $\dfrac{56}{98} = \text{—}$

(4) $\dfrac{35}{70} = \text{—}$ (5) $\dfrac{6}{36} = \text{—}$ (6) $\dfrac{60}{80} = \text{—}$

(7) $\dfrac{8}{72} = \text{—}$ (8) $\dfrac{20}{80} = \text{—}$ (9) $\dfrac{24}{30} = \text{—}$

(10) $\dfrac{24}{96} = \text{—}$ (11) $\dfrac{8}{80} = \text{—}$ (12) $\dfrac{9}{63} = \text{—}$

(13) $\dfrac{32}{48} = \text{—}$ (14) $\dfrac{50}{80} = \text{—}$ (15) $\dfrac{15}{30} = \text{—}$

(16) $\dfrac{36}{48} = \text{—}$ (17) $\dfrac{63}{72} = \text{—}$ (18) $\dfrac{10}{50} = \text{—}$

(19) $\dfrac{27}{54} = \text{—}$ (20) $\dfrac{64}{96} = \text{—}$ (21) $\dfrac{36}{48} = \text{—}$

(22) $\dfrac{63}{84} = \text{—}$ (23) $\dfrac{44}{88} = \text{—}$ (24) $\dfrac{54}{99} = \text{—}$

(25) $\dfrac{20}{80} = \text{—}$ (26) $\dfrac{10}{70} = \text{—}$ (27) $\dfrac{12}{36} = \text{—}$

(28) $\dfrac{60}{80} = \text{—}$ (29) $\dfrac{24}{84} = \text{—}$ (30) $\dfrac{25}{40} = \text{—}$

(31) $\dfrac{21}{35} = \text{—}$ (32) $\dfrac{16}{96} = \text{—}$ (33) $\dfrac{35}{105} = \text{—}$

(34) $\dfrac{40}{60} = \text{—}$ (35) $\dfrac{18}{54} = \text{—}$ (36) $\dfrac{20}{25} = \text{—}$

(37) $\dfrac{70}{90} = \text{—}$ (38) $\dfrac{12}{48} = \text{—}$ (39) $\dfrac{12}{15} = \text{—}$

(40) $\dfrac{16}{64} = \text{—}$ (41) $\dfrac{20}{100} = \text{—}$ (42) $\dfrac{72}{80} = \text{—}$

(43) $\dfrac{45}{75} = \text{—}$ (44) $\dfrac{28}{56} = \text{—}$ (45) $\dfrac{35}{56} = \text{—}$

Reduce the fractions to the simplest form :

(1) $\dfrac{28}{56} = \underline{\quad}$

(2) $\dfrac{72}{80} = \underline{\quad}$

(3) $\dfrac{70}{80} = \underline{\quad}$

(4) $\dfrac{48}{64} = \underline{\quad}$

(5) $\dfrac{45}{50} = \underline{\quad}$

(6) $\dfrac{14}{28} = \underline{\quad}$

(7) $\dfrac{81}{90} = \underline{\quad}$

(8) $\dfrac{42}{70} = \underline{\quad}$

(9) $\dfrac{50}{90} = \underline{\quad}$

(10) $\dfrac{56}{84} = \underline{\quad}$

(11) $\dfrac{72}{90} = \underline{\quad}$

(12) $\dfrac{40}{60} = \underline{\quad}$

(13) $\dfrac{12}{96} = \underline{\quad}$

(14) $\dfrac{54}{90} = \underline{\quad}$

(15) $\dfrac{80}{96} = \underline{\quad}$

(16) $\dfrac{9}{45} = \underline{\quad}$

(17) $\dfrac{16}{32} = \underline{\quad}$

(18) $\dfrac{28}{56} = \underline{\quad}$

(19) $\dfrac{32}{64} = \underline{\quad}$

(20) $\dfrac{32}{96} = \underline{\quad}$

(21) $\dfrac{10}{30} = \underline{\quad}$

(22) $\dfrac{36}{72} = \underline{\quad}$

(23) $\dfrac{15}{60} = \underline{\quad}$

(24) $\dfrac{24}{60} = \underline{\quad}$

(25) $\dfrac{30}{50} = \underline{\quad}$

(26) $\dfrac{35}{50} = \underline{\quad}$

(27) $\dfrac{27}{63} = \underline{\quad}$

(28) $\dfrac{45}{90} = \underline{\quad}$

(29) $\dfrac{24}{40} = \underline{\quad}$

(30) $\dfrac{72}{90} = \underline{\quad}$

(31) $\dfrac{25}{100} = \underline{\quad}$

(32) $\dfrac{20}{50} = \underline{\quad}$

(33) $\dfrac{45}{90} = \underline{\quad}$

(34) $\dfrac{30}{90} = \underline{\quad}$

(35) $\dfrac{56}{98} = \underline{\quad}$

(36) $\dfrac{7}{21} = \underline{\quad}$

(37) $\dfrac{44}{88} = \underline{\quad}$

(38) $\dfrac{25}{75} = \underline{\quad}$

(39) $\dfrac{18}{36} = \underline{\quad}$

(40) $\dfrac{72}{90} = \underline{\quad}$

(41) $\dfrac{48}{64} = \underline{\quad}$

(42) $\dfrac{12}{60} = \underline{\quad}$

(43) $\dfrac{21}{28} = \underline{\quad}$

(44) $\dfrac{56}{72} = \underline{\quad}$

(45) $\dfrac{16}{64} = \underline{\quad}$

Find the equivalent fractions:

(1) $\dfrac{4}{11} = \dfrac{16}{}$

(2) $\dfrac{1}{11} = \dfrac{}{33}$

(3) $\dfrac{1}{10} = \dfrac{}{30}$

(4) $\dfrac{1}{9} = \dfrac{}{27}$

(5) $\dfrac{5}{6} = \dfrac{}{60}$

(6) $\dfrac{2}{3} = \dfrac{}{12}$

(7) $\dfrac{1}{6} = \dfrac{}{6}$

(8) $\dfrac{4}{9} = \dfrac{}{36}$

(9) $\dfrac{3}{8} = \dfrac{}{32}$

(10) $\dfrac{2}{7} = \dfrac{}{28}$

(11) $\dfrac{1}{2} = \dfrac{14}{}$

(12) $\dfrac{2}{6} = \dfrac{}{12}$

(13) $\dfrac{2}{11} = \dfrac{}{22}$

(14) $\dfrac{4}{7} = \dfrac{}{28}$

(15) $\dfrac{4}{7} = \dfrac{}{28}$

(16) $\dfrac{1}{8} = \dfrac{}{8}$

(17) $\dfrac{1}{5} = \dfrac{5}{}$

(18) $\dfrac{3}{5} = \dfrac{}{15}$

(19) $\dfrac{4}{6} = \dfrac{}{24}$

(20) $\dfrac{2}{10} = \dfrac{}{20}$

(21) $\dfrac{7}{10} = \dfrac{}{70}$

(22) $\dfrac{7}{11} = \dfrac{}{77}$

(23) $\dfrac{2}{8} = \dfrac{}{16}$

(24) $\dfrac{5}{9} = \dfrac{}{45}$

(25) $\dfrac{3}{10} = \dfrac{}{60}$

(26) $\dfrac{5}{9} = \dfrac{}{45}$

(27) $\dfrac{1}{4} = \dfrac{4}{}$

(28) $\dfrac{5}{6} = \dfrac{20}{}$

(29) $\dfrac{1}{2} = \dfrac{}{16}$

(30) $\dfrac{3}{10} = \dfrac{}{30}$

(31) $\dfrac{7}{12} = \dfrac{14}{}$

(32) $\dfrac{2}{7} = \dfrac{}{21}$

(33) $\dfrac{7}{12} = \dfrac{}{84}$

(34) $\dfrac{1}{5} = \dfrac{}{25}$

(35) $\dfrac{3}{8} = \dfrac{}{48}$

(36) $\dfrac{6}{7} = \dfrac{}{42}$

(37) $\dfrac{7}{12} = \dfrac{}{84}$

(38) $\dfrac{3}{6} = \dfrac{}{18}$

(39) $\dfrac{1}{4} = \dfrac{}{3}$

(40) $\dfrac{2}{3} = \dfrac{6}{}$

(41) $\dfrac{2}{7} = \dfrac{}{14}$

(42) $\dfrac{7}{8} = \dfrac{}{56}$

(43) $\dfrac{2}{5} = \dfrac{}{20}$

(44) $\dfrac{1}{3} = \dfrac{2}{}$

(45) $\dfrac{2}{5} = \dfrac{6}{}$

Find the equivalent fractions:

(1) $\dfrac{7}{12} = \dfrac{14}{}$

(2) $\dfrac{3}{11} = \dfrac{}{33}$

(3) $\dfrac{3}{5} = \dfrac{}{20}$

(4) $\dfrac{1}{9} = \dfrac{}{9}$

(5) $\dfrac{1}{4} = \dfrac{12}{}$

(6) $\dfrac{3}{7} = \dfrac{36}{}$

(7) $\dfrac{1}{12} = \dfrac{}{12}$

(8) $\dfrac{1}{11} = \dfrac{}{33}$

(9) $\dfrac{2}{7} = \dfrac{}{14}$

(10) $\dfrac{5}{12} = \dfrac{}{60}$

(11) $\dfrac{4}{5} = \dfrac{36}{}$

(12) $\dfrac{4}{7} = \dfrac{16}{}$

(13) $\dfrac{2}{12} = \dfrac{}{48}$

(14) $\dfrac{2}{9} = \dfrac{}{36}$

(15) $\dfrac{5}{6} = \dfrac{}{30}$

(16) $\dfrac{2}{4} = \dfrac{}{8}$

(17) $\dfrac{7}{8} = \dfrac{}{56}$

(18) $\dfrac{1}{6} = \dfrac{2}{}$

(19) $\dfrac{6}{7} = \dfrac{42}{}$

(20) $\dfrac{4}{6} = \dfrac{}{24}$

(21) $\dfrac{1}{4} = \dfrac{}{12}$

(22) $\dfrac{1}{3} = \dfrac{}{15}$

(23) $\dfrac{4}{7} = \dfrac{20}{}$

(24) $\dfrac{3}{8} = \dfrac{}{32}$

(25) $\dfrac{4}{7} = \dfrac{}{28}$

(26) $\dfrac{4}{5} = \dfrac{72}{}$

(27) $\dfrac{2}{3} = \dfrac{}{18}$

(28) $\dfrac{3}{8} = \dfrac{24}{}$

(29) $\dfrac{3}{7} = \dfrac{36}{}$

(30) $\dfrac{4}{9} = \dfrac{}{36}$

(31) $\dfrac{1}{2} = \dfrac{4}{}$

(32) $\dfrac{5}{6} = \dfrac{}{30}$

(33) $\dfrac{7}{10} = \dfrac{}{70}$

(34) $\dfrac{1}{8} = \dfrac{}{16}$

(35) $\dfrac{1}{3} = \dfrac{}{27}$

(36) $\dfrac{3}{11} = \dfrac{}{33}$

(37) $\dfrac{5}{8} = \dfrac{}{80}$

(38) $\dfrac{3}{4} = \dfrac{}{24}$

(39) $\dfrac{7}{9} = \dfrac{28}{}$

(40) $\dfrac{1}{5} = \dfrac{7}{}$

(41) $\dfrac{5}{6} = \dfrac{}{30}$

(42) $\dfrac{2}{5} = \dfrac{}{30}$

(43) $\dfrac{3}{12} = \dfrac{}{36}$

(44) $\dfrac{7}{12} = \dfrac{}{84}$

(45) $\dfrac{3}{10} = \dfrac{}{30}$

Find the equivalent fractions:

(1) $\dfrac{2}{3} = \dfrac{4}{}$

(2) $\dfrac{1}{8} = \dfrac{}{16}$

(3) $\dfrac{1}{12} = \dfrac{}{12}$

(4) $\dfrac{1}{10} = \dfrac{2}{}$

(5) $\dfrac{2}{3} = \dfrac{}{18}$

(6) $\dfrac{5}{6} = \dfrac{}{30}$

(7) $\dfrac{1}{6} = \dfrac{}{18}$

(8) $\dfrac{2}{9} = \dfrac{}{18}$

(9) $\dfrac{5}{15} = \dfrac{30}{}$

(10) $\dfrac{7}{12} = \dfrac{}{84}$

(11) $\dfrac{1}{2} = \dfrac{10}{}$

(12) $\dfrac{1}{10} = \dfrac{2}{}$

(13) $\dfrac{1}{2} = \dfrac{}{24}$

(14) $\dfrac{2}{11} = \dfrac{}{22}$

(15) $\dfrac{1}{2} = \dfrac{}{24}$

(16) $\dfrac{5}{12} = \dfrac{}{60}$

(17) $\dfrac{3}{9} = \dfrac{}{27}$

(18) $\dfrac{1}{4} = \dfrac{}{12}$

(19) $\dfrac{5}{9} = \dfrac{}{45}$

(20) $\dfrac{7}{10} = \dfrac{}{70}$

(21) $\dfrac{2}{3} = \dfrac{}{12}$

(22) $\dfrac{6}{11} = \dfrac{}{66}$

(23) $\dfrac{6}{7} = \dfrac{}{42}$

(24) $\dfrac{6}{7} = \dfrac{}{42}$

(25) $\dfrac{3}{6} = \dfrac{}{18}$

(26) $\dfrac{2}{3} = \dfrac{12}{}$

(27) $\dfrac{3}{5} = \dfrac{}{15}$

(28) $\dfrac{7}{9} = \dfrac{}{63}$

(29) $\dfrac{1}{9} = \dfrac{}{27}$

(30) $\dfrac{2}{5} = \dfrac{16}{}$

(31) $\dfrac{2}{9} = \dfrac{}{18}$

(32) $\dfrac{4}{7} = \dfrac{20}{}$

(33) $\dfrac{7}{11} = \dfrac{}{77}$

(34) $\dfrac{5}{7} = \dfrac{}{35}$

(35) $\dfrac{2}{8} = \dfrac{}{16}$

(36) $\dfrac{7}{8} = \dfrac{}{56}$

(37) $\dfrac{3}{9} = \dfrac{}{27}$

(38) $\dfrac{7}{9} = \dfrac{}{63}$

(39) $\dfrac{1}{5} = \dfrac{}{25}$

(40) $\dfrac{1}{5} = \dfrac{}{40}$

(41) $\dfrac{3}{12} = \dfrac{}{36}$

(42) $\dfrac{4}{7} = \dfrac{20}{}$

(43) $\dfrac{1}{4} = \dfrac{2}{}$

(44) $\dfrac{1}{11} = \dfrac{}{11}$

(45) $\dfrac{2}{9} = \dfrac{}{18}$

Find the equivalent fractions:

(1) $\dfrac{1}{8} = \dfrac{}{16}$

(2) $\dfrac{5}{8} = \dfrac{}{40}$

(3) $\dfrac{5}{7} = \dfrac{}{28}$

(4) $\dfrac{7}{8} = \dfrac{}{56}$

(5) $\dfrac{5}{8} = \dfrac{20}{}$

(6) $\dfrac{1}{3} = \dfrac{}{9}$

(7) $\dfrac{2}{6} = \dfrac{}{12}$

(8) $\dfrac{3}{8} = \dfrac{6}{}$

(9) $\dfrac{1}{10} = \dfrac{2}{}$

(10) $\dfrac{1}{2} = \dfrac{}{10}$

(11) $\dfrac{3}{12} = \dfrac{}{36}$

(12) $\dfrac{1}{5} = \dfrac{5}{}$

(13) $\dfrac{3}{10} = \dfrac{}{30}$

(14) $\dfrac{2}{3} = \dfrac{}{18}$

(15) $\dfrac{1}{12} = \dfrac{}{12}$

(16) $\dfrac{2}{5} = \dfrac{16}{}$

(17) $\dfrac{3}{5} = \dfrac{}{15}$

(18) $\dfrac{1}{3} = \dfrac{6}{}$

(19) $\dfrac{3}{8} = \dfrac{}{24}$

(20) $\dfrac{1}{3} = \dfrac{6}{}$

(21) $\dfrac{2}{6} = \dfrac{}{12}$

(22) $\dfrac{1}{2} = \dfrac{3}{}$

(23) $\dfrac{1}{9} = \dfrac{}{9}$

(24) $\dfrac{4}{9} = \dfrac{}{36}$

(25) $\dfrac{1}{10} = \dfrac{2}{}$

(26) $\dfrac{5}{7} = \dfrac{}{28}$

(27) $\dfrac{2}{3} = \dfrac{}{18}$

(28) $\dfrac{2}{3} = \dfrac{12}{}$

(29) $\dfrac{1}{3} = \dfrac{}{9}$

(30) $\dfrac{2}{9} = \dfrac{}{36}$

(31) $\dfrac{3}{7} = \dfrac{36}{}$

(32) $\dfrac{2}{5} = \dfrac{}{50}$

(33) $\dfrac{1}{9} = \dfrac{}{9}$

(34) $\dfrac{5}{6} = \dfrac{}{42}$

(35) $\dfrac{1}{4} = \dfrac{6}{}$

(36) $\dfrac{2}{3} = \dfrac{8}{}$

(37) $\dfrac{4}{9} = \dfrac{20}{}$

(38) $\dfrac{1}{10} = \dfrac{}{30}$

(39) $\dfrac{2}{4} = \dfrac{}{8}$

(40) $\dfrac{1}{3} = \dfrac{5}{}$

(41) $\dfrac{1}{3} = \dfrac{8}{}$

(42) $\dfrac{4}{11} = \dfrac{16}{}$

(43) $\dfrac{3}{5} = \dfrac{}{30}$

(44) $\dfrac{1}{11} = \dfrac{}{33}$

(45) $\dfrac{1}{2} = \dfrac{}{24}$

Find the equivalent fractions:

(1) $\dfrac{3}{4} = \dfrac{}{16}$

(2) $\dfrac{1}{4} = \dfrac{}{28}$

(3) $\dfrac{5}{10} = \dfrac{25}{}$

(4) $\dfrac{2}{5} = \dfrac{}{10}$

(5) $\dfrac{4}{5} = \dfrac{16}{}$

(6) $\dfrac{5}{11} = \dfrac{}{55}$

(7) $\dfrac{5}{6} = \dfrac{}{30}$

(8) $\dfrac{1}{12} = \dfrac{}{12}$

(9) $\dfrac{2}{5} = \dfrac{20}{}$

(10) $\dfrac{2}{6} = \dfrac{}{12}$

(11) $\dfrac{1}{4} = \dfrac{5}{}$

(12) $\dfrac{1}{2} = \dfrac{}{16}$

(13) $\dfrac{4}{11} = \dfrac{}{44}$

(14) $\dfrac{4}{7} = \dfrac{}{28}$

(15) $\dfrac{3}{5} = \dfrac{}{15}$

(16) $\dfrac{1}{6} = \dfrac{}{12}$

(17) $\dfrac{1}{4} = \dfrac{3}{}$

(18) $\dfrac{3}{9} = \dfrac{}{27}$

(19) $\dfrac{2}{12} = \dfrac{}{72}$

(20) $\dfrac{7}{11} = \dfrac{63}{}$

(21) $\dfrac{1}{8} = \dfrac{}{16}$

(22) $\dfrac{5}{8} = \dfrac{40}{}$

(23) $\dfrac{7}{8} = \dfrac{}{56}$

(24) $\dfrac{2}{5} = \dfrac{}{30}$

(25) $\dfrac{3}{6} = \dfrac{}{18}$

(26) $\dfrac{4}{7} = \dfrac{}{56}$

(27) $\dfrac{1}{5} = \dfrac{}{25}$

(28) $\dfrac{2}{9} = \dfrac{}{18}$

(29) $\dfrac{1}{2} = \dfrac{17}{}$

(30) $\dfrac{2}{6} = \dfrac{}{12}$

(31) $\dfrac{2}{3} = \dfrac{}{18}$

(32) $\dfrac{2}{3} = \dfrac{}{24}$

(33) $\dfrac{3}{11} = \dfrac{}{33}$

(34) $\dfrac{1}{10} = \dfrac{}{10}$

(35) $\dfrac{1}{12} = \dfrac{}{12}$

(36) $\dfrac{7}{9} = \dfrac{}{63}$

(37) $\dfrac{5}{8} = \dfrac{}{40}$

(38) $\dfrac{2}{12} = \dfrac{}{24}$

(39) $\dfrac{1}{9} = \dfrac{}{27}$

(40) $\dfrac{3}{5} = \dfrac{}{20}$

(41) $\dfrac{7}{9} = \dfrac{}{63}$

(42) $\dfrac{2}{6} = \dfrac{}{36}$

(43) $\dfrac{7}{8} = \dfrac{}{56}$

(44) $\dfrac{5}{6} = \dfrac{}{30}$

(45) $\dfrac{7}{12} = \dfrac{}{84}$

Find the equivalent fractions:

(1) $\dfrac{1}{6} = \dfrac{}{18}$

(2) $\dfrac{1}{7} = \dfrac{}{21}$

(3) $\dfrac{3}{6} = \dfrac{}{18}$

(4) $\dfrac{1}{6} = \dfrac{}{18}$

(5) $\dfrac{7}{10} = \dfrac{14}{}$

(6) $\dfrac{1}{11} = \dfrac{}{11}$

(7) $\dfrac{1}{3} = \dfrac{}{9}$

(8) $\dfrac{7}{12} = \dfrac{}{84}$

(9) $\dfrac{1}{3} = \dfrac{}{12}$

(10) $\dfrac{1}{4} = \dfrac{6}{}$

(11) $\dfrac{1}{8} = \dfrac{}{16}$

(12) $\dfrac{2}{8} = \dfrac{}{64}$

(13) $\dfrac{1}{3} = \dfrac{}{9}$

(14) $\dfrac{5}{12} = \dfrac{}{60}$

(15) $\dfrac{2}{4} = \dfrac{}{24}$

(16) $\dfrac{2}{4} = \dfrac{}{16}$

(17) $\dfrac{2}{9} = \dfrac{}{36}$

(18) $\dfrac{4}{9} = \dfrac{}{27}$

(19) $\dfrac{3}{4} = \dfrac{}{16}$

(20) $\dfrac{2}{3} = \dfrac{6}{}$

(21) $\dfrac{5}{7} = \dfrac{15}{}$

(22) $\dfrac{4}{9} = \dfrac{20}{}$

(23) $\dfrac{3}{6} = \dfrac{}{18}$

(24) $\dfrac{1}{5} = \dfrac{}{5}$

(25) $\dfrac{4}{8} = \dfrac{}{32}$

(26) $\dfrac{4}{7} = \dfrac{16}{}$

(27) $\dfrac{6}{11} = \dfrac{}{66}$

(28) $\dfrac{6}{7} = \dfrac{}{42}$

(29) $\dfrac{2}{8} = \dfrac{}{16}$

(30) $\dfrac{1}{11} = \dfrac{}{11}$

(31) $\dfrac{4}{11} = \dfrac{}{44}$

(32) $\dfrac{5}{7} = \dfrac{}{35}$

(33) $\dfrac{2}{6} = \dfrac{}{18}$

(34) $\dfrac{1}{3} = \dfrac{4}{}$

(35) $\dfrac{4}{5} = \dfrac{8}{}$

(36) $\dfrac{3}{5} = \dfrac{}{25}$

(37) $\dfrac{1}{12} = \dfrac{}{48}$

(38) $\dfrac{3}{7} = \dfrac{}{21}$

(39) $\dfrac{3}{7} = \dfrac{}{21}$

(40) $\dfrac{2}{11} = \dfrac{}{22}$

(41) $\dfrac{1}{10} = \dfrac{}{10}$

(42) $\dfrac{2}{12} = \dfrac{}{24}$

(43) $\dfrac{1}{12} = \dfrac{}{12}$

(44) $\dfrac{1}{7} = \dfrac{}{7}$

(45) $\dfrac{4}{6} = \dfrac{}{24}$

Find the equivalent fractions:

(1) $\dfrac{4}{8} = \dfrac{}{32}$ (2) $\dfrac{2}{10} = \dfrac{}{20}$ (3) $\dfrac{3}{4} = \dfrac{15}{}$

(4) $\dfrac{4}{5} = \dfrac{20}{}$ (5) $\dfrac{5}{8} = \dfrac{}{40}$ (6) $\dfrac{3}{10} = \dfrac{15}{}$

(7) $\dfrac{5}{6} = \dfrac{50}{}$ (8) $\dfrac{2}{3} = \dfrac{6}{}$ (9) $\dfrac{5}{12} = \dfrac{}{60}$

(10) $\dfrac{3}{8} = \dfrac{24}{}$ (11) $\dfrac{2}{11} = \dfrac{}{22}$ (12) $\dfrac{7}{8} = \dfrac{}{56}$

(13) $\dfrac{4}{5} = \dfrac{}{20}$ (14) $\dfrac{1}{12} = \dfrac{}{12}$ (15) $\dfrac{4}{9} = \dfrac{}{72}$

(16) $\dfrac{2}{7} = \dfrac{8}{}$ (17) $\dfrac{5}{9} = \dfrac{}{45}$ (18) $\dfrac{7}{12} = \dfrac{}{84}$

(19) $\dfrac{2}{3} = \dfrac{8}{}$ (20) $\dfrac{4}{9} = \dfrac{}{36}$ (21) $\dfrac{3}{11} = \dfrac{}{33}$

(22) $\dfrac{5}{12} = \dfrac{10}{}$ (23) $\dfrac{5}{12} = \dfrac{}{60}$ (24) $\dfrac{7}{9} = \dfrac{28}{}$

(25) $\dfrac{3}{6} = \dfrac{}{18}$ (26) $\dfrac{1}{5} = \dfrac{3}{}$ (27) $\dfrac{4}{11} = \dfrac{}{44}$

(28) $\dfrac{1}{6} = \dfrac{}{18}$ (29) $\dfrac{1}{7} = \dfrac{}{14}$ (30) $\dfrac{1}{6} = \dfrac{}{6}$

(31) $\dfrac{2}{9} = \dfrac{}{18}$ (32) $\dfrac{1}{7} = \dfrac{}{14}$ (33) $\dfrac{1}{4} = \dfrac{25}{}$

(34) $\dfrac{4}{11} = \dfrac{}{44}$ (35) $\dfrac{1}{4} = \dfrac{}{4}$ (36) $\dfrac{1}{3} = \dfrac{}{9}$

(37) $\dfrac{7}{10} = \dfrac{}{70}$ (38) $\dfrac{5}{6} = \dfrac{}{30}$ (39) $\dfrac{3}{8} = \dfrac{}{48}$

(40) $\dfrac{4}{7} = \dfrac{}{28}$ (41) $\dfrac{2}{3} = \dfrac{}{12}$ (42) $\dfrac{3}{4} = \dfrac{}{24}$

(43) $\dfrac{2}{7} = \dfrac{16}{}$ (44) $\dfrac{4}{9} = \dfrac{}{27}$ (45) $\dfrac{2}{9} = \dfrac{}{18}$

Find the equivalent fractions:

(1) $\dfrac{3}{4} = \dfrac{6}{}$

(2) $\dfrac{2}{11} = \dfrac{}{22}$

(3) $\dfrac{1}{4} = \dfrac{}{32}$

(4) $\dfrac{3}{4} = \dfrac{}{16}$

(5) $\dfrac{5}{11} = \dfrac{}{55}$

(6) $\dfrac{7}{11} = \dfrac{63}{}$

(7) $\dfrac{2}{5} = \dfrac{}{50}$

(8) $\dfrac{2}{3} = \dfrac{12}{}$

(9) $\dfrac{5}{6} = \dfrac{25}{}$

(10) $\dfrac{5}{6} = \dfrac{10}{}$

(11) $\dfrac{7}{10} = \dfrac{}{70}$

(12) $\dfrac{3}{4} = \dfrac{24}{}$

(13) $\dfrac{5}{6} = \dfrac{}{60}$

(14) $\dfrac{1}{8} = \dfrac{}{32}$

(15) $\dfrac{3}{10} = \dfrac{15}{}$

(16) $\dfrac{2}{5} = \dfrac{}{50}$

(17) $\dfrac{5}{9} = \dfrac{}{45}$

(18) $\dfrac{1}{2} = \dfrac{}{24}$

(19) $\dfrac{3}{8} = \dfrac{}{24}$

(20) $\dfrac{7}{11} = \dfrac{}{77}$

(21) $\dfrac{3}{5} = \dfrac{}{15}$

(22) $\dfrac{2}{9} = \dfrac{}{54}$

(23) $\dfrac{1}{9} = \dfrac{}{9}$

(24) $\dfrac{1}{2} = \dfrac{40}{}$

(25) $\dfrac{4}{5} = \dfrac{24}{}$

(26) $\dfrac{3}{5} = \dfrac{}{25}$

(27) $\dfrac{1}{6} = \dfrac{}{18}$

(28) $\dfrac{3}{10} = \dfrac{}{30}$

(29) $\dfrac{4}{11} = \dfrac{}{44}$

(30) $\dfrac{1}{5} = \dfrac{20}{}$

(31) $\dfrac{3}{5} = \dfrac{}{15}$

(32) $\dfrac{3}{7} = \dfrac{}{21}$

(33) $\dfrac{5}{8} = \dfrac{}{40}$

(34) $\dfrac{2}{11} = \dfrac{}{22}$

(35) $\dfrac{3}{9} = \dfrac{}{27}$

(36) $\dfrac{1}{2} = \dfrac{}{8}$

(37) $\dfrac{3}{9} = \dfrac{}{27}$

(38) $\dfrac{3}{7} = \dfrac{}{21}$

(39) $\dfrac{3}{8} = \dfrac{}{24}$

(40) $\dfrac{2}{5} = \dfrac{}{30}$

(41) $\dfrac{4}{5} = \dfrac{20}{}$

(42) $\dfrac{3}{12} = \dfrac{}{36}$

(43) $\dfrac{7}{9} = \dfrac{}{63}$

(44) $\dfrac{1}{3} = \dfrac{6}{}$

(45) $\dfrac{2}{7} = \dfrac{8}{}$

Find the equivalent fractions:

(1) $\dfrac{3}{11} = \dfrac{}{33}$

(2) $\dfrac{1}{5} = \dfrac{5}{}$

(3) $\dfrac{3}{11} = \dfrac{}{33}$

(4) $\dfrac{4}{7} = \dfrac{}{28}$

(5) $\dfrac{5}{8} = \dfrac{40}{}$

(6) $\dfrac{4}{9} = \dfrac{}{36}$

(7) $\dfrac{3}{10} = \dfrac{}{30}$

(8) $\dfrac{1}{8} = \dfrac{}{16}$

(9) $\dfrac{2}{7} = \dfrac{8}{}$

(10) $\dfrac{1}{4} = \dfrac{6}{}$

(11) $\dfrac{3}{5} = \dfrac{}{30}$

(12) $\dfrac{2}{12} = \dfrac{}{24}$

(13) $\dfrac{1}{8} = \dfrac{3}{}$

(14) $\dfrac{7}{11} = \dfrac{21}{}$

(15) $\dfrac{6}{7} = \dfrac{36}{}$

(16) $\dfrac{3}{8} = \dfrac{}{48}$

(17) $\dfrac{3}{11} = \dfrac{}{33}$

(18) $\dfrac{2}{7} = \dfrac{}{14}$

(19) $\dfrac{4}{7} = \dfrac{}{28}$

(20) $\dfrac{3}{12} = \dfrac{}{36}$

(21) $\dfrac{2}{8} = \dfrac{}{40}$

(22) $\dfrac{2}{5} = \dfrac{}{20}$

(23) $\dfrac{2}{7} = \dfrac{}{14}$

(24) $\dfrac{3}{9} = \dfrac{}{27}$

(25) $\dfrac{5}{12} = \dfrac{}{60}$

(26) $\dfrac{1}{11} = \dfrac{}{11}$

(27) $\dfrac{5}{10} = \dfrac{25}{}$

(28) $\dfrac{4}{7} = \dfrac{}{28}$

(29) $\dfrac{2}{3} = \dfrac{14}{}$

(30) $\dfrac{4}{6} = \dfrac{}{24}$

(31) $\dfrac{2}{3} = \dfrac{8}{}$

(32) $\dfrac{1}{12} = \dfrac{}{12}$

(33) $\dfrac{3}{4} = \dfrac{24}{}$

(34) $\dfrac{3}{5} = \dfrac{}{25}$

(35) $\dfrac{5}{6} = \dfrac{30}{}$

(36) $\dfrac{1}{5} = \dfrac{5}{}$

(37) $\dfrac{7}{11} = \dfrac{63}{}$

(38) $\dfrac{2}{3} = \dfrac{}{18}$

(39) $\dfrac{2}{9} = \dfrac{}{18}$

(40) $\dfrac{1}{3} = \dfrac{}{27}$

(41) $\dfrac{5}{10} = \dfrac{25}{}$

(42) $\dfrac{5}{9} = \dfrac{}{45}$

(43) $\dfrac{1}{2} = \dfrac{12}{}$

(44) $\dfrac{4}{5} = \dfrac{}{40}$

(45) $\dfrac{1}{9} = \dfrac{}{27}$

Find the equivalent fractions:

(1) $\dfrac{2}{5} = \dfrac{}{20}$

(2) $\dfrac{5}{8} = \dfrac{40}{}$

(3) $\dfrac{3}{8} = \dfrac{24}{}$

(4) $\dfrac{2}{9} = \dfrac{}{18}$

(5) $\dfrac{1}{7} = \dfrac{}{14}$

(6) $\dfrac{3}{11} = \dfrac{}{33}$

(7) $\dfrac{3}{9} = \dfrac{}{27}$

(8) $\dfrac{3}{8} = \dfrac{24}{}$

(9) $\dfrac{5}{8} = \dfrac{40}{}$

(10) $\dfrac{1}{7} = \dfrac{}{21}$

(11) $\dfrac{5}{9} = \dfrac{}{45}$

(12) $\dfrac{5}{6} = \dfrac{30}{}$

(13) $\dfrac{7}{10} = \dfrac{}{70}$

(14) $\dfrac{7}{10} = \dfrac{}{70}$

(15) $\dfrac{6}{7} = \dfrac{42}{}$

(16) $\dfrac{1}{4} = \dfrac{}{8}$

(17) $\dfrac{5}{7} = \dfrac{35}{}$

(18) $\dfrac{5}{8} = \dfrac{20}{}$

(19) $\dfrac{4}{5} = \dfrac{8}{}$

(20) $\dfrac{1}{7} = \dfrac{}{21}$

(21) $\dfrac{3}{4} = \dfrac{24}{}$

(22) $\dfrac{3}{12} = \dfrac{}{24}$

(23) $\dfrac{5}{10} = \dfrac{25}{}$

(24) $\dfrac{3}{11} = \dfrac{}{33}$

(25) $\dfrac{7}{12} = \dfrac{}{84}$

(26) $\dfrac{1}{2} = \dfrac{6}{}$

(27) $\dfrac{5}{8} = \dfrac{}{80}$

(28) $\dfrac{3}{9} = \dfrac{}{27}$

(29) $\dfrac{2}{8} = \dfrac{}{16}$

(30) $\dfrac{1}{4} = \dfrac{}{3}$

(31) $\dfrac{1}{6} = \dfrac{}{18}$

(32) $\dfrac{1}{2} = \dfrac{6}{}$

(33) $\dfrac{2}{8} = \dfrac{}{32}$

(34) $\dfrac{3}{6} = \dfrac{}{24}$

(35) $\dfrac{5}{8} = \dfrac{10}{}$

(36) $\dfrac{5}{8} = \dfrac{}{40}$

(37) $\dfrac{1}{9} = \dfrac{}{27}$

(38) $\dfrac{1}{7} = \dfrac{}{21}$

(39) $\dfrac{5}{6} = \dfrac{}{30}$

(40) $\dfrac{5}{10} = \dfrac{15}{}$

(41) $\dfrac{1}{2} = \dfrac{}{10}$

(42) $\dfrac{1}{5} = \dfrac{20}{}$

(43) $\dfrac{5}{11} = \dfrac{}{55}$

(44) $\dfrac{3}{4} = \dfrac{9}{}$

(45) $\dfrac{5}{7} = \dfrac{35}{}$

Day: 1 Date: Score: /45
Name: Time: : Rating: ☆☆☆☆☆

Convert decimals to fractions in their simplest form:

(1) 0.4 = —
(2) 1.125 = —
(3) 1.9 = —
(4) 0.125 = —
(5) 9.5 = —
(6) 6.2 = —
(7) 0.875 = —
(8) 3.625 = —
(9) 3.4 = —
(10) 0.3 = —
(11) 0.166 = —
(12) 8.5 = —
(13) 0.6 = —
(14) 5.8 = —
(15) 5.3 = —
(16) 0.2 = —
(17) 2.25 = —
(18) 2.1 = —
(19) 0.666 = —
(20) 4.666 = —
(21) 7.7 = —
(22) 0.9 = —
(23) 1.8 = —
(24) 4.1 = —
(25) 0.25 = —
(26) 8.125 = —
(27) 9.6 = —
(28) 0.625 = —
(29) 3.2 = —
(30) 6.9 = —
(31) 0.1 = —
(32) 0.4 = —
(33) 1.7 = —
(34) 0.125 = —
(35) 6.375 = —
(36) 3.8 = —
(37) 0.375 = —
(38) 2.125 = —
(39) 8.2 = —
(40) 0.875 = —
(41) 5.5 = —
(42) 5.9 = —
(43) 0.625 = —
(44) 1.333 = —
(45) 2.6 = —

Convert decimals to fractions in their simplest form:

(1) $0.5 = \overline{}$

(2) $7.25 = \overline{}$

(3) $7.3 = \overline{}$

(4) $0.333 = \overline{}$

(5) $4.8 = \overline{}$

(6) $4.8 = \overline{}$

(7) $0.6 = \overline{}$

(8) $0.125 = \overline{}$

(9) $9.8 = \overline{}$

(10) $0.125 = \overline{}$

(11) $3.333 = \overline{}$

(12) $5.2 = \overline{}$

(13) $0.2 = \overline{}$

(14) $2.8 = \overline{}$

(15) $1.3 = \overline{}$

(16) $0.625 = \overline{}$

(17) $6.125 = \overline{}$

(18) $3.2 = \overline{}$

(19) $0.375 = \overline{}$

(20) $1.625 = \overline{}$

(21) $8.8 = \overline{}$

(22) $0.9 = \overline{}$

(23) $8.375 = \overline{}$

(24) $6.5 = \overline{}$

(25) $0.5 = \overline{}$

(26) $4.125 = \overline{}$

(27) $2.9 = \overline{}$

(28) $0.625 = \overline{}$

(29) $5.666 = \overline{}$

(30) $7.2 = \overline{}$

(31) $0.25 = \overline{}$

(32) $2.375 = \overline{}$

(33) $4.9 = \overline{}$

(34) $0.6 = \overline{}$

(35) $7.2 = \overline{}$

(36) $9.7 = \overline{}$

(37) $0.75 = \overline{}$

(38) $2.4 = \overline{}$

(39) $5.1 = \overline{}$

(40) $0.125 = \overline{}$

(41) $0.3 = \overline{}$

(42) $1.4 = \overline{}$

(43) $0.375 = \overline{}$

(44) $3.6 = \overline{}$

(45) $3.6 = \overline{}$

Day: 3 | Date: | Score: /45

Name: | Time: : | Rating: ☆☆☆☆☆

Convert decimals to fractions in their simplest form:

(1) 0.875 = —	(2) 1.6 = —	(3) 8.9 = —
(4) 0.625 = —	(5) 2.2 = —	(6) 6.4 = —
(7) 0.125 = —	(8) 0.9 = —	(9) 2.3 = —
(10) 0.5 = —	(11) 1.3 = —	(12) 7.8 = —
(13) 0.625 = —	(14) 3.3 = —	(15) 4.5 = —
(16) 0.25 = —	(17) 1.7 = —	(18) 9.5 = —
(19) 0.75 = —	(20) 2.5 = —	(21) 5.7 = —
(22) 0.125 = —	(23) 0.6 = —	(24) 1.2 = —
(25) 0.375 = —	(26) 3.8 = —	(27) 3.7 = —
(28) 0.875 = —	(29) 1.2 = —	(30) 8.7 = —
(31) 2.5 = —	(32) 2.1 = —	(33) 6.3 = —
(34) 1.7 = —	(35) 0.4 = —	(36) 2.8 = —
(37) 2.5 = —	(38) 3.5 = —	(39) 7.6 = —
(40) 3.9 = —	(41) 1.5 = —	(42) 4.2 = —
(43) 6.25 = —	(44) 2.9 = —	(45) 9.3 = —

Day:	4	Date:		Score:	/45
Name:		Time:	:	Rating:	☆☆☆☆☆

Convert decimals to fractions in their simplest form:

(1) 2.75 = —	(2) 0.8 = —	(3) 2.75 = —
(4) 1.5 = —	(5) 3.4 = —	(6) 3.6 = —
(7) 3.2 = —	(8) 1.4 = —	(9) 4.25 = —
(10) 7.125 = —	(11) 2.3 = —	(12) 5.125 = —
(13) 0.75 = —	(14) 0.2 = —	(15) 6.4 = —
(16) 9.625 = —	(17) 3.1 = —	(18) 7.75 = —
(19) 4.333 = —	(20) 1.1 = —	(21) 8.333 = —
(22) 5.5 = —	(23) 2.6 = —	(24) 9.875 = —
(25) 2.875 = —	(26) 0.7 = —	(27) 10.2 = —
(28) 0.375 = —	(29) 3.7 = —	(30) 1.3 = —
(31) 6.8 = —	(32) 1.8 = —	(33) 2.5 = —
(34) 1.75 = —	(35) 2.8 = —	(36) 3.2 = —
(37) 2.6 = —	(38) 0.5 = —	(39) 4.4 = —
(40) 4.125 = —	(41) 3.2 = —	(42) 5.7 = —
(43) 0.8 = —	(44) 1.9 = —	(45) 6.1 = —

Day: 5 Date: Score: /45

Name: Time: : Rating: ☆☆☆☆☆

Convert decimals to fractions in their simplest form:

(1) 8.625 = —	(2) 2.7 = —	(3) 7.8 = —
(4) 3.2 = —	(5) 0.1 = —	(6) 8.9 = —
(7) 5.25 = —	(8) 3.9 = —	(9) 1.6 = —
(10) 2.2 = —	(11) 4.6 = —	(12) 2.9 = —
(13) 0.625 = —	(14) 8.4 = —	(15) 3.5 = —
(16) 7.5 = —	(17) 5.5 = —	(18) 4.1 = —
(19) 1.6 = —	(20) 9.5 = —	(21) 5.3 = —
(22) 3.375 = —	(23) 6.8 = —	(24) 6.7 = —
(25) 0.6 = —	(26) 2.9 = —	(27) 7.2 = —
(28) 9.75 = —	(29) 7.2 = —	(30) 8.5 = —
(31) 2.333 = —	(32) 1.6 = —	(33) 1.9 = —
(34) 6.625 = —	(35) 3.3 = —	(36) 2.7 = —
(37) 1.8 = —	(38) 8.6 = —	(39) 3.6 = —
(40) 4.2 = —	(41) 6.3 = —	(42) 4.3 = —
(43) 0.875 = —	(44) 2.2 = —	(45) 5.6 = —

Convert decimals to fractions in their simplest form:

(1) $5.5 = —$

(2) $7.9 = —$

(3) $6.4 = —$

(4) $7.375 = —$

(5) $4.2 = —$

(6) $7.1 = —$

(7) $1.25 = —$

(8) $9.1 = —$

(9) $8.8 = —$

(10) $2.875 = —$

(11) $5.7 = —$

(12) $1.4 = —$

(13) $2.125 = —$

(14) $1.2 = —$

(15) $2.2 = —$

(16) $3.4 = —$

(17) $3.7 = —$

(18) $3.9 = —$

(19) $0.25 = —$

(20) $8.3 = —$

(21) $4.5 = —$

(22) $4.8 = —$

(23) $6.1 = —$

(24) $5.2 = —$

(25) $1.625 = —$

(26) $2.8 = —$

(27) $6.3 = —$

(28) $2.25 = —$

(29) $7.1 = —$

(30) $7.6 = —$

(31) $7.2 = —$

(32) $4.9 = —$

(33) $8.2 = —$

(34) $5.5 = —$

(35) $9.7 = —$

(36) $1.7 = —$

(37) $0.5 = —$

(38) $5.1 = —$

(39) $2.3 = —$

(40) $3.6 = —$

(41) $1.4 = —$

(42) $3.8 = —$

(43) $6.125 = —$

(44) $3.6 = —$

(45) $4.9 = —$

Convert decimals to fractions in their simplest form:

(1) $0.875 = \underline{}$

(2) $8.9 = \underline{}$

(3) $5.4 = \underline{}$

(4) $1.4 = \underline{}$

(5) $6.4 = \underline{}$

(6) $6.9 = \underline{}$

(7) $8.25 = \underline{}$

(8) $2.3 = \underline{}$

(9) $7.3 = \underline{}$

(10) $2.8 = \underline{}$

(11) $7.8 = \underline{}$

(12) $8.6 = \underline{}$

(13) $5.875 = \underline{}$

(14) $4.4 = \underline{}$

(15) $1.2 = \underline{}$

(16) $1.125 = \underline{}$

(17) $9.2 = \underline{}$

(18) $2.4 = \underline{}$

(19) $7.5 = \underline{}$

(20) $5.8 = \underline{}$

(21) $3.4 = \underline{}$

(22) $2.166 = \underline{}$

(23) $1.1 = \underline{}$

(24) $4.8 = \underline{}$

(25) $3.75 = \underline{}$

(26) $3.5 = \underline{}$

(27) $5.5 = \underline{}$

(28) $1.2 = \underline{}$

(29) $8.7 = \underline{}$

(30) $6.6 = \underline{}$

(31) $6.625 = \underline{}$

(32) $6.5 = \underline{}$

(33) $7.9 = \underline{}$

(34) $0.333 = \underline{}$

(35) $2.5 = \underline{}$

(36) $8.3 = \underline{}$

(37) $4.5 = \underline{}$

(38) $7.5 = \underline{}$

(39) $1.8 = \underline{}$

(40) $9.8 = \underline{}$

(41) $4.5 = \underline{}$

(42) $2.8 = \underline{}$

(43) $2.125 = \underline{}$

(44) $2.4 = \underline{}$

(45) $1.7 = \underline{}$

Day: 8

Name:

Date:

Time: :

Score: /45

Rating: ☆☆☆☆☆

Convert decimals to fractions in their simplest form:

(1) $0.625 = \underline{\quad}$

(2) $7.4 = \underline{\quad}$

(3) $2.3 = \underline{\quad}$

(4) $5.4 = \underline{\quad}$

(5) $4.7 = \underline{\quad}$

(6) $3.9 = \underline{\quad}$

(7) $1.875 = \underline{\quad}$

(8) $9.4 = \underline{\quad}$

(9) $4.2 = \underline{\quad}$

(10) $3.2 = \underline{\quad}$

(11) $5.6 = \underline{\quad}$

(12) $5.4 = \underline{\quad}$

(13) $6.375 = \underline{\quad}$

(14) $1.9 = \underline{\quad}$

(15) $6.9 = \underline{\quad}$

(16) $0.875 = \underline{\quad}$

(17) $6.2 = \underline{\quad}$

(18) $7.3 = \underline{\quad}$

(19) $2.2 = \underline{\quad}$

(20) $3.4 = \underline{\quad}$

(21) $8.6 = \underline{\quad}$

(22) $4.25 = \underline{\quad}$

(23) $8.5 = \underline{\quad}$

(24) $1.5 = \underline{\quad}$

(25) $1.333 = \underline{\quad}$

(26) $5.3 = \underline{\quad}$

(27) $2.8 = \underline{\quad}$

(28) $8.5 = \underline{\quad}$

(29) $2.1 = \underline{\quad}$

(30) $3.4 = \underline{\quad}$

(31) $3.625 = \underline{\quad}$

(32) $7.7 = \underline{\quad}$

(33) $4.8 = \underline{\quad}$

(34) $0.2 = \underline{\quad}$

(35) $4.1 = \underline{\quad}$

(36) $5.5 = \underline{\quad}$

(37) $6.8 = \underline{\quad}$

(38) $9.6 = \underline{\quad}$

(39) $6.6 = \underline{\quad}$

(40) $2.625 = \underline{\quad}$

(41) $6.9 = \underline{\quad}$

(42) $7.9 = \underline{\quad}$

(43) $5.2 = \underline{\quad}$

(44) $1.7 = \underline{\quad}$

(45) $8.3 = \underline{\quad}$

Day: 9	Date:	Score: /45
Name:	Time: :	Rating: ☆☆☆☆☆

Convert decimals to fractions in their simplest form:

(1) 1.125 = —	(2) 3.8 = —	(3) 1.9 = —
(4) 4.666 = —	(5) 8.2 = —	(6) 2.7 = —
(7) 0.75 = —	(8) 5.9 = —	(9) 3.6 = —
(10) 7.25 = —	(11) 2.6 = —	(12) 4.3 = —
(13) 3.8 = —	(14) 7.3 = —	(15) 5.6 = —
(16) 2.375 = —	(17) 4.8 = —	(18) 6.4 = —
(19) 5.625 = —	(20) 9.8 = —	(21) 7.1 = —
(22) 1.6 = —	(23) 5.2 = —	(24) 8.8 = —
(25) 2.375 = —	(26) 1.3 = —	(27) 1.4 = —
(28) 0.8 = —	(29) 3.2 = —	(30) 2.2 = —
(31) 3.125 = —	(32) 8.8 = —	(33) 3.8 = —
(34) 5.4 = —	(35) 6.5 = —	(36) 4.9 = —
(37) 1.2 = —	(38) 2.9 = —	(39) 5.2 = —
(40) 7.625 = —	(41) 7.2 = —	(42) 6.3 = —
(43) 4.5 = —	(44) 4.9 = —	(45) 7.6 = —

Convert decimals to fractions in their simplest form:

(1) $0.875 = $ —

(2) $9.7 = $ —

(3) $8.2 = $ —

(4) $2.2 = $ —

(5) $5.1 = $ —

(6) $1.6 = $ —

(7) $6.75 = $ —

(8) $1.4 = $ —

(9) $2.9 = $ —

(10) $1.625 = $ —

(11) $3.6 = $ —

(12) $3.5 = $ —

(13) $8.2 = $ —

(14) $8.9 = $ —

(15) $4.1 = $ —

(16) $3.8 = $ —

(17) $6.4 = $ —

(18) $5.3 = $ —

(19) $0.25 = $ —

(20) $2.3 = $ —

(21) $6.7 = $ —

(22) $5.125 = $ —

(23) $7.8 = $ —

(24) $7.2 = $ —

(25) $2.666 = $ —

(26) $4.5 = $ —

(27) $8.5 = $ —

(28) $4.25 = $ —

(29) $9.5 = $ —

(30) $1.3 = $ —

(31) $1.4 = $ —

(32) $2.4 = $ —

(33) $2.5 = $ —

(34) $7.875 = $ —

(35) $7.4 = $ —

(36) $3.2 = $ —

(37) $0.6 = $ —

(38) $4.7 = $ —

(39) $4.4 = $ —

(40) $2.875 = $ —

(41) $9.4 = $ —

(42) $5.7 = $ —

(43) $6.4 = $ —

(44) $5.6 = $ —

(45) $6.1 = $ —

Adding fractions with like denominators:

(1) $\dfrac{3}{5} + \dfrac{2}{5} = -$

(2) $\dfrac{5}{7} + \dfrac{5}{7} = -$

(3) $\dfrac{3}{7} + \dfrac{1}{7} = -$

(4) $\dfrac{2}{6} + \dfrac{1}{6} = -$

(5) $\dfrac{2}{8} + \dfrac{2}{8} = -$

(6) $\dfrac{2}{9} + \dfrac{2}{9} = -$

(7) $\dfrac{5}{9} + \dfrac{4}{9} = -$

(8) $\dfrac{3}{10} + \dfrac{3}{10} = -$

(9) $\dfrac{1}{12} + \dfrac{3}{12} = -$

(10) $\dfrac{4}{12} + \dfrac{7}{12} = -$

(11) $\dfrac{4}{12} + \dfrac{4}{12} = -$

(12) $\dfrac{5}{8} + \dfrac{3}{8} = -$

(13) $\dfrac{2}{3} + \dfrac{1}{3} = -$

(14) $\dfrac{1}{5} + \dfrac{1}{5} = -$

(15) $\dfrac{2}{8} + \dfrac{2}{8} = -$

(16) $\dfrac{1}{2} + \dfrac{1}{2} = -$

(17) $\dfrac{5}{9} + \dfrac{5}{9} = -$

(18) $\dfrac{1}{8} + \dfrac{2}{8} = -$

(19) $\dfrac{7}{10} + \dfrac{4}{10} = -$

(20) $\dfrac{2}{9} + \dfrac{2}{9} = -$

(21) $\dfrac{4}{8} + \dfrac{4}{8} = -$

(22) $\dfrac{5}{8} + \dfrac{2}{8} = -$

(23) $\dfrac{3}{12} + \dfrac{3}{12} = -$

(24) $\dfrac{3}{9} + \dfrac{4}{9} = -$

(25) $\dfrac{2}{7} + \dfrac{5}{7} = -$

(26) $\dfrac{4}{6} + \dfrac{4}{6} = -$

(27) $\dfrac{1}{3} + \dfrac{3}{3} = -$

(28) $\dfrac{3}{4} + \dfrac{1}{4} = -$

(29) $\dfrac{1}{12} + \dfrac{1}{12} = -$

(30) $\dfrac{2}{4} + \dfrac{3}{4} = -$

(31) $\dfrac{2}{5} + \dfrac{3}{5} = -$

(32) $\dfrac{5}{10} + \dfrac{5}{10} = -$

(33) $\dfrac{4}{5} + \dfrac{2}{5} = -$

(34) $\dfrac{4}{7} + \dfrac{2}{7} = -$

(35) $\dfrac{2}{12} + \dfrac{2}{12} = -$

(36) $\dfrac{3}{7} + \dfrac{5}{7} = -$

(37) $\dfrac{3}{8} + \dfrac{5}{8} = -$

(38) $\dfrac{3}{7} + \dfrac{3}{7} = -$

(39) $\dfrac{1}{6} + \dfrac{1}{6} = -$

(40) $\dfrac{1}{3} + \dfrac{1}{3} = -$

(41) $\dfrac{4}{8} + \dfrac{4}{8} = -$

(42) $\dfrac{2}{7} + \dfrac{5}{7} = -$

(43) $\dfrac{5}{6} + \dfrac{4}{6} = -$

(44) $\dfrac{1}{7} + \dfrac{1}{7} = -$

(45) $\dfrac{4}{12} + \dfrac{4}{12} = -$

Adding fractions with like denominators:

(1) $\dfrac{2}{4} + \dfrac{3}{4} = $ —

(2) $\dfrac{5}{6} + \dfrac{5}{6} = $ —

(3) $\dfrac{3}{4} + \dfrac{4}{4} = $ —

(4) $\dfrac{7}{8} + \dfrac{2}{8} = $ —

(5) $\dfrac{2}{10} + \dfrac{2}{10} = $ —

(6) $\dfrac{1}{8} + \dfrac{1}{8} = $ —

(7) $\dfrac{5}{9} + \dfrac{1}{9} = $ —

(8) $\dfrac{3}{12} + \dfrac{3}{12} = $ —

(9) $\dfrac{2}{8} + \dfrac{2}{8} = $ —

(10) $\dfrac{3}{10} + \dfrac{5}{10} = $ —

(11) $\dfrac{4}{9} + \dfrac{4}{9} = $ —

(12) $\dfrac{5}{9} + \dfrac{1}{9} = $ —

(13) $\dfrac{1}{5} + \dfrac{3}{5} = $ —

(14) $\dfrac{1}{4} + \dfrac{1}{4} = $ —

(15) $\dfrac{1}{5} + \dfrac{2}{5} = $ —

(16) $\dfrac{2}{6} + \dfrac{4}{6} = $ —

(17) $\dfrac{5}{8} + \dfrac{5}{8} = $ —

(18) $\dfrac{3}{10} + \dfrac{5}{10} = $ —

(19) $\dfrac{5}{10} + \dfrac{2}{10} = $ —

(20) $\dfrac{2}{5} + \dfrac{2}{5} = $ —

(21) $\dfrac{2}{3} + \dfrac{3}{3} = $ —

(22) $\dfrac{1}{2} + \dfrac{1}{2} = $ —

(23) $\dfrac{3}{6} + \dfrac{3}{6} = $ —

(24) $\dfrac{4}{7} + \dfrac{1}{7} = $ —

(25) $\dfrac{4}{12} + \dfrac{7}{12} = $ —

(26) $\dfrac{4}{7} + \dfrac{4}{7} = $ —

(27) $\dfrac{1}{9} + \dfrac{2}{9} = $ —

(28) $\dfrac{3}{7} + \dfrac{2}{7} = $ —

(29) $\dfrac{1}{10} + \dfrac{1}{10} = $ —

(30) $\dfrac{3}{6} + \dfrac{2}{6} = $ —

(31) $\dfrac{1}{6} + \dfrac{3}{6} = $ —

(32) $\dfrac{5}{12} + \dfrac{5}{12} = $ —

(33) $\dfrac{2}{12} + \dfrac{2}{12} = $ —

(34) $\dfrac{5}{10} + \dfrac{3}{10} = $ —

(35) $\dfrac{2}{7} + \dfrac{2}{7} = $ —

(36) $\dfrac{4}{10} + \dfrac{4}{10} = $ —

(37) $\dfrac{2}{3} + \dfrac{1}{3} = $ —

(38) $\dfrac{3}{9} + \dfrac{3}{9} = $ —

(39) $\dfrac{1}{6} + \dfrac{2}{6} = $ —

(40) $\dfrac{7}{9} + \dfrac{4}{9} = $ —

(41) $\dfrac{4}{12} + \dfrac{4}{12} = $ —

(42) $\dfrac{1}{4} + \dfrac{3}{4} = $ —

(43) $\dfrac{4}{8} + \dfrac{3}{8} = $ —

(44) $\dfrac{1}{4} + \dfrac{1}{4} = $ —

(45) $\dfrac{2}{9} + \dfrac{4}{9} = $ —

Adding fractions with like denominators:

(1) $\dfrac{2}{10} + \dfrac{4}{10} = \text{—}$

(2) $\dfrac{5}{9} + \dfrac{5}{9} = \text{—}$

(3) $\dfrac{2}{3} + \dfrac{3}{3} = \text{—}$

(4) $\dfrac{3}{4} + \dfrac{1}{4} = \text{—}$

(5) $\dfrac{2}{6} + \dfrac{2}{6} = \text{—}$

(6) $\dfrac{1}{7} + \dfrac{6}{7} = \text{—}$

(7) $\dfrac{1}{8} + \dfrac{3}{8} = \text{—}$

(8) $\dfrac{3}{8} + \dfrac{3}{8} = \text{—}$

(9) $\dfrac{5}{8} + \dfrac{6}{8} = \text{—}$

(10) $\dfrac{6}{10} + \dfrac{2}{10} = \text{—}$

(11) $\dfrac{4}{10} + \dfrac{4}{10} = \text{—}$

(12) $\dfrac{3}{8} + \dfrac{5}{8} = \text{—}$

(13) $\dfrac{2}{7} + \dfrac{1}{7} = \text{—}$

(14) $\dfrac{1}{5} + \dfrac{1}{5} = \text{—}$

(15) $\dfrac{1}{8} + \dfrac{8}{8} = \text{—}$

(16) $\dfrac{4}{6} + \dfrac{2}{6} = \text{—}$

(17) $\dfrac{1}{6} + \dfrac{4}{6} = \text{—}$

(18) $\dfrac{1}{5} + \dfrac{5}{5} = \text{—}$

(19) $\dfrac{5}{12} + \dfrac{4}{12} = \text{—}$

(20) $\dfrac{1}{4} + \dfrac{3}{4} = \text{—}$

(21) $\dfrac{3}{5} + \dfrac{5}{5} = \text{—}$

(22) $\dfrac{1}{9} + \dfrac{2}{9} = \text{—}$

(23) $\dfrac{1}{8} + \dfrac{2}{8} = \text{—}$

(24) $\dfrac{1}{12} + \dfrac{6}{12} = \text{—}$

(25) $\dfrac{3}{5} + \dfrac{2}{5} = \text{—}$

(26) $\dfrac{1}{10} + \dfrac{2}{10} = \text{—}$

(27) $\dfrac{5}{6} + \dfrac{6}{6} = \text{—}$

(28) $\dfrac{6}{7} + \dfrac{1}{7} = \text{—}$

(29) $\dfrac{2}{9} + \dfrac{4}{9} = \text{—}$

(30) $\dfrac{3}{10} + \dfrac{6}{10} = \text{—}$

(31) $\dfrac{2}{8} + \dfrac{1}{8} = \text{—}$

(32) $\dfrac{1}{7} + \dfrac{5}{7} = \text{—}$

(33) $\dfrac{4}{5} + \dfrac{5}{5} = \text{—}$

(34) $\dfrac{4}{10} + \dfrac{6}{10} = \text{—}$

(35) $\dfrac{3}{8} + \dfrac{5}{8} = \text{—}$

(36) $\dfrac{1}{8} + \dfrac{4}{8} = \text{—}$

(37) $\dfrac{1}{4} + \dfrac{3}{4} = \text{—}$

(38) $\dfrac{2}{12} + \dfrac{3}{12} = \text{—}$

(39) $\dfrac{4}{9} + \dfrac{9}{9} = \text{—}$

(40) $\dfrac{5}{9} + \dfrac{3}{9} = \text{—}$

(41) $\dfrac{1}{12} + \dfrac{5}{12} = \text{—}$

(42) $\dfrac{2}{4} + \dfrac{4}{4} = \text{—}$

(43) $\dfrac{2}{10} + \dfrac{1}{10} = \text{—}$

(44) $\dfrac{3}{12} + \dfrac{5}{12} = \text{—}$

(45) $\dfrac{1}{10} + \dfrac{4}{10} = \text{—}$

Adding fractions with like denominators:

(1) $\dfrac{3}{6} + \dfrac{2}{6} = $ ——

(2) $\dfrac{2}{5} + \dfrac{3}{5} = $ ——

(3) $\dfrac{2}{12} + \dfrac{4}{12} = $ ——

(4) $\dfrac{4}{5} + \dfrac{1}{5} = $ ——

(5) $\dfrac{1}{9} + \dfrac{5}{9} = $ ——

(6) $\dfrac{1}{3} + \dfrac{4}{3} = $ ——

(7) $\dfrac{7}{12} + \dfrac{4}{12} = $ ——

(8) $\dfrac{2}{7} + \dfrac{5}{7} = $ ——

(9) $\dfrac{3}{7} + \dfrac{6}{7} = $ ——

(10) $\dfrac{1}{7} + \dfrac{4}{7} = $ ——

(11) $\dfrac{1}{12} + \dfrac{4}{12} = $ ——

(12) $\dfrac{2}{8} + \dfrac{8}{8} = $ ——

(13) $\dfrac{2}{9} + \dfrac{4}{9} = $ ——

(14) $\dfrac{2}{4} + \dfrac{3}{4} = $ ——

(15) $\dfrac{1}{9} + \dfrac{3}{9} = $ ——

(16) $\dfrac{3}{10} + \dfrac{2}{10} = $ ——

(17) $\dfrac{1}{10} + \dfrac{4}{10} = $ ——

(18) $\dfrac{1}{9} + \dfrac{3}{9} = $ ——

(19) $\dfrac{4}{10} + \dfrac{3}{10} = $ ——

(20) $\dfrac{2}{6} + \dfrac{4}{6} = $ ——

(21) $\dfrac{3}{8} + \dfrac{3}{8} = $ ——

(22) $\dfrac{5}{7} + \dfrac{3}{7} = $ ——

(23) $\dfrac{3}{10} + \dfrac{4}{10} = $ ——

(24) $\dfrac{2}{10} + \dfrac{7}{10} = $ ——

(25) $\dfrac{1}{3} + \dfrac{2}{3} = $ ——

(26) $\dfrac{1}{5} + \dfrac{4}{5} = $ ——

(27) $\dfrac{1}{12} + \dfrac{10}{12} = $ ——

(28) $\dfrac{2}{4} + \dfrac{1}{4} = $ ——

(29) $\dfrac{2}{8} + \dfrac{4}{8} = $ ——

(30) $\dfrac{1}{8} + \dfrac{8}{8} = $ ——

(31) $\dfrac{3}{8} + \dfrac{2}{8} = $ ——

(32) $\dfrac{1}{7} + \dfrac{4}{7} = $ ——

(33) $\dfrac{3}{6} + \dfrac{3}{6} = $ ——

(34) $\dfrac{4}{7} + \dfrac{1}{7} = $ ——

(35) $\dfrac{2}{12} + \dfrac{4}{12} = $ ——

(36) $\dfrac{1}{3} + \dfrac{3}{3} = $ ——

(37) $\dfrac{5}{10} + \dfrac{13}{10} = $ ——

(38) $\dfrac{3}{9} + \dfrac{5}{9} = $ ——

(39) $\dfrac{2}{10} + \dfrac{8}{10} = $ ——

(40) $\dfrac{1}{5} + \dfrac{2}{5} = $ ——

(41) $\dfrac{1}{8} + \dfrac{5}{8} = $ ——

(42) $\dfrac{4}{7} + \dfrac{4}{7} = $ ——

(43) $\dfrac{6}{9} + \dfrac{3}{9} = $ ——

(44) $\dfrac{2}{3} + \dfrac{5}{3} = $ ——

(45) $\dfrac{1}{12} + \dfrac{6}{12} = $ ——

Adding fractions with like denominators:

(1) $\dfrac{3}{6} + \dfrac{2}{6} = $ —

(2) $\dfrac{3}{5} + \dfrac{5}{5} = $ —

(3) $\dfrac{4}{8} + \dfrac{6}{8} = $ —

(4) $\dfrac{1}{9} + \dfrac{3}{9} = $ —

(5) $\dfrac{2}{10} + \dfrac{3}{10} = $ —

(6) $\dfrac{1}{4} + \dfrac{6}{4} = $ —

(7) $\dfrac{2}{12} + \dfrac{5}{12} = $ —

(8) $\dfrac{1}{3} + \dfrac{5}{3} = $ —

(9) $\dfrac{2}{7} + \dfrac{3}{7} = $ —

(10) $\dfrac{1}{2} + \dfrac{1}{2} = $ —

(11) $\dfrac{2}{9} + \dfrac{5}{9} = $ —

(12) $\dfrac{2}{6} + \dfrac{4}{6} = $ —

(13) $\dfrac{7}{8} + \dfrac{3}{8} = $ —

(14) $\dfrac{1}{6} + \dfrac{3}{6} = $ —

(15) $\dfrac{1}{10} + \dfrac{8}{10} = $ —

(16) $\dfrac{4}{6} + \dfrac{1}{6} = $ —

(17) $\dfrac{2}{12} + \dfrac{4}{12} = $ —

(18) $\dfrac{4}{10} + \dfrac{4}{10} = $ —

(19) $\dfrac{5}{7} + \dfrac{4}{7} = $ —

(20) $\dfrac{1}{12} + \dfrac{3}{12} = $ —

(21) $\dfrac{3}{12} + \dfrac{7}{12} = $ —

(22) $\dfrac{2}{5} + \dfrac{4}{5} = $ —

(23) $\dfrac{2}{8} + \dfrac{3}{8} = $ —

(24) $\dfrac{1}{6} + \dfrac{3}{6} = $ —

(25) $\dfrac{1}{4} + \dfrac{3}{4} = $ —

(26) $\dfrac{1}{4} + \dfrac{5}{4} = $ —

(27) $\dfrac{3}{4} + \dfrac{7}{4} = $ —

(28) $\dfrac{3}{9} + \dfrac{6}{9} = $ —

(29) $\dfrac{3}{7} + \dfrac{5}{7} = $ —

(30) $\dfrac{2}{5} + \dfrac{3}{5} = $ —

(31) $\dfrac{2}{7} + \dfrac{4}{7} = $ —

(32) $\dfrac{1}{9} + \dfrac{2}{9} = $ —

(33) $\dfrac{3}{9} + \dfrac{9}{9} = $ —

(34) $\dfrac{1}{3} + \dfrac{2}{3} = $ —

(35) $\dfrac{2}{6} + \dfrac{5}{6} = $ —

(36) $\dfrac{1}{12} + \dfrac{11}{12} = $ —

(37) $\dfrac{5}{10} + \dfrac{4}{10} = $ —

(38) $\dfrac{3}{12} + \dfrac{5}{12} = $ —

(39) $\dfrac{2}{8} + \dfrac{3}{8} = $ —

(40) $\dfrac{3}{5} + \dfrac{1}{5} = $ —

(41) $\dfrac{1}{5} + \dfrac{5}{5} = $ —

(42) $\dfrac{1}{9} + \dfrac{9}{9} = $ —

(43) $\dfrac{7}{12} + \dfrac{5}{12} = $ —

(44) $\dfrac{2}{7} + \dfrac{3}{7} = $ —

(45) $\dfrac{1}{5} + \dfrac{7}{5} = $ —

Adding fractions with like denominators:

(1) $\dfrac{2}{6} + \dfrac{3}{6} = $ —

(2) $\dfrac{1}{12} + \dfrac{5}{12} = $ —

(3) $\dfrac{2}{9} + \dfrac{9}{9} = $ —

(4) $\dfrac{5}{10} + \dfrac{4}{10} = $ —

(5) $\dfrac{3}{6} + \dfrac{5}{6} = $ —

(6) $\dfrac{1}{8} + \dfrac{3}{8} = $ —

(7) $\dfrac{1}{5} + \dfrac{1}{5} = $ —

(8) $\dfrac{2}{4} + \dfrac{5}{4} = $ —

(9) $\dfrac{2}{12} + \dfrac{3}{12} = $ —

(10) $\dfrac{4}{8} + \dfrac{2}{8} = $ —

(11) $\dfrac{3}{10} + \dfrac{5}{10} = $ —

(12) $\dfrac{3}{5} + \dfrac{7}{5} = $ —

(13) $\dfrac{3}{7} + \dfrac{3}{7} = $ —

(14) $\dfrac{1}{8} + \dfrac{4}{8} = $ —

(15) $\dfrac{1}{4} + \dfrac{5}{4} = $ —

(16) $\dfrac{2}{8} + \dfrac{2}{8} = $ —

(17) $\dfrac{2}{5} + \dfrac{4}{5} = $ —

(18) $\dfrac{1}{8} + \dfrac{7}{8} = $ —

(19) $\dfrac{5}{9} + \dfrac{5}{9} = $ —

(20) $\dfrac{1}{10} + \dfrac{5}{10} = $ —

(21) $\dfrac{2}{4} + \dfrac{3}{4} = $ —

(22) $\dfrac{1}{7} + \dfrac{5}{7} = $ —

(23) $\dfrac{2}{12} + \dfrac{3}{12} = $ —

(24) $\dfrac{4}{5} + \dfrac{7}{5} = $ —

(25) $\dfrac{3}{4} + \dfrac{3}{4} = $ —

(26) $\dfrac{1}{7} + \dfrac{3}{7} = $ —

(27) $\dfrac{1}{10} + \dfrac{10}{10} = $ —

(28) $\dfrac{4}{7} + \dfrac{5}{7} = $ —

(29) $\dfrac{2}{3} + \dfrac{4}{3} = $ —

(30) $\dfrac{4}{6} + \dfrac{5}{6} = $ —

(31) $\dfrac{2}{4} + \dfrac{2}{4} = $ —

(32) $\dfrac{3}{5} + \dfrac{4}{5} = $ —

(33) $\dfrac{2}{7} + \dfrac{5}{7} = $ —

(34) $\dfrac{1}{6} + \dfrac{2}{6} = $ —

(35) $\dfrac{1}{12} + \dfrac{4}{12} = $ —

(36) $\dfrac{5}{7} + \dfrac{2}{7} = $ —

(37) $\dfrac{3}{10} + \dfrac{4}{10} = $ —

(38) $\dfrac{3}{8} + \dfrac{4}{8} = $ —

(39) $\dfrac{2}{9} + \dfrac{5}{9} = $ —

(40) $\dfrac{5}{6} + \dfrac{3}{6} = $ —

(41) $\dfrac{2}{12} + \dfrac{5}{12} = $ —

(42) $\dfrac{1}{8} + \dfrac{7}{8} = $ —

(43) $\dfrac{1}{8} + \dfrac{4}{8} = $ —

(44) $\dfrac{1}{4} + \dfrac{4}{4} = $ —

(45) $\dfrac{3}{6} + \dfrac{4}{6} = $ —

Adding fractions with like denominators:

(1) $\dfrac{7}{9} + \dfrac{2}{9} =$ ___

(2) $\dfrac{2}{10} + \dfrac{4}{10} =$ ___

(3) $\dfrac{4}{10} + \dfrac{7}{10} =$ ___

(4) $\dfrac{2}{10} + \dfrac{2}{10} =$ ___

(5) $\dfrac{1}{6} + \dfrac{5}{6} =$ ___

(6) $\dfrac{2}{10} + \dfrac{4}{10} =$ ___

(7) $\dfrac{4}{10} + \dfrac{6}{10} =$ ___

(8) $\dfrac{3}{9} + \dfrac{4}{9} =$ ___

(9) $\dfrac{3}{10} + \dfrac{8}{10} =$ ___

(10) $\dfrac{3}{8} + \dfrac{3}{8} =$ ___

(11) $\dfrac{2}{8} + \dfrac{5}{8} =$ ___

(12) $\dfrac{1}{12} + \dfrac{4}{12} =$ ___

(13) $\dfrac{1}{3} + \dfrac{1}{3} =$ ___

(14) $\dfrac{1}{5} + \dfrac{3}{5} =$ ___

(15) $\dfrac{2}{10} + \dfrac{3}{10} =$ ___

(16) $\dfrac{2}{3} + \dfrac{2}{3} =$ ___

(17) $\dfrac{2}{8} + \dfrac{2}{8} =$ ___

(18) $\dfrac{2}{12} + \dfrac{9}{12} =$ ___

(19) $\dfrac{5}{12} + \dfrac{3}{12} =$ ___

(20) $\dfrac{3}{10} + \dfrac{3}{10} =$ ___

(21) $\dfrac{1}{7} + \dfrac{4}{7} =$ ___

(22) $\dfrac{4}{9} + \dfrac{1}{9} =$ ___

(23) $\dfrac{2}{5} + \dfrac{3}{5} =$ ___

(24) $\dfrac{3}{8} + \dfrac{5}{8} =$ ___

(25) $\dfrac{1}{10} + \dfrac{2}{10} =$ ___

(26) $\dfrac{1}{9} + \dfrac{7}{9} =$ ___

(27) $\dfrac{1}{10} + \dfrac{7}{10} =$ ___

(28) $\dfrac{3}{7} + \dfrac{4}{7} =$ ___

(29) $\dfrac{3}{12} + \dfrac{4}{12} =$ ___

(30) $\dfrac{2}{5} + \dfrac{3}{5} =$ ___

(31) $\dfrac{2}{9} + \dfrac{3}{9} =$ ___

(32) $\dfrac{2}{8} + \dfrac{7}{8} =$ ___

(33) $\dfrac{1}{9} + \dfrac{8}{9} =$ ___

(34) $\dfrac{4}{10} + \dfrac{5}{10} =$ ___

(35) $\dfrac{1}{6} + \dfrac{5}{6} =$ ___

(36) $\dfrac{2}{3} + \dfrac{3}{3} =$ ___

(37) $\dfrac{1}{4} + \dfrac{2}{4} =$ ___

(38) $\dfrac{3}{10} + \dfrac{7}{10} =$ ___

(39) $\dfrac{1}{5} + \dfrac{3}{5} =$ ___

(40) $\dfrac{3}{5} + \dfrac{3}{5} =$ ___

(41) $\dfrac{2}{7} + \dfrac{5}{7} =$ ___

(42) $\dfrac{1}{10} + \dfrac{9}{10} =$ ___

(43) $\dfrac{5}{8} + \dfrac{4}{8} =$ ___

(44) $\dfrac{1}{8} + \dfrac{5}{8} =$ ___

(45) $\dfrac{2}{8} + \dfrac{6}{8} =$ ___

Adding fractions with like denominators:

(1) $\dfrac{2}{10} + \dfrac{3}{10} = \underline{}$

(2) $\dfrac{2}{12} + \dfrac{5}{12} = \underline{}$

(3) $\dfrac{1}{3} + \dfrac{2}{3} = \underline{}$

(4) $\dfrac{1}{7} + \dfrac{3}{7} = \underline{}$

(5) $\dfrac{1}{5} + \dfrac{4}{5} = \underline{}$

(6) $\dfrac{2}{7} + \dfrac{3}{7} = \underline{}$

(7) $\dfrac{6}{10} + \dfrac{4}{10} = \underline{}$

(8) $\dfrac{3}{4} + \dfrac{5}{4} = \underline{}$

(9) $\dfrac{3}{12} + \dfrac{5}{12} = \underline{}$

(10) $\dfrac{2}{5} + \dfrac{3}{5} = \underline{}$

(11) $\dfrac{2}{12} + \dfrac{5}{12} = \underline{}$

(12) $\dfrac{4}{8} + \dfrac{6}{8} = \underline{}$

(13) $\dfrac{4}{12} + \dfrac{6}{12} = \underline{}$

(14) $\dfrac{1}{7} + \dfrac{4}{7} = \underline{}$

(15) $\dfrac{1}{4} + \dfrac{3}{4} = \underline{}$

(16) $\dfrac{3}{9} + \dfrac{6}{9} = \underline{}$

(17) $\dfrac{3}{6} + \dfrac{4}{6} = \underline{}$

(18) $\dfrac{2}{4} + \dfrac{4}{4} = \underline{}$

(19) $\dfrac{4}{10} + \dfrac{6}{10} = \underline{}$

(20) $\dfrac{2}{10} + \dfrac{3}{10} = \underline{}$

(21) $\dfrac{2}{6} + \dfrac{4}{6} = \underline{}$

(22) $\dfrac{1}{4} + \dfrac{4}{4} = \underline{}$

(23) $\dfrac{1}{12} + \dfrac{4}{12} = \underline{}$

(24) $\dfrac{1}{2} + \dfrac{3}{2} = \underline{}$

(25) $\dfrac{2}{7} + \dfrac{2}{7} = \underline{}$

(26) $\dfrac{2}{4} + \dfrac{6}{4} = \underline{}$

(27) $\dfrac{3}{5} + \dfrac{4}{5} = \underline{}$

(28) $\dfrac{1}{2} + \dfrac{3}{2} = \underline{}$

(29) $\dfrac{1}{9} + \dfrac{6}{9} = \underline{}$

(30) $\dfrac{1}{6} + \dfrac{5}{6} = \underline{}$

(31) $\dfrac{2}{10} + \dfrac{5}{10} = \underline{}$

(32) $\dfrac{3}{7} + \dfrac{4}{7} = \underline{}$

(33) $\dfrac{1}{12} + \dfrac{6}{12} = \underline{}$

(34) $\dfrac{1}{8} + \dfrac{1}{8} = \underline{}$

(35) $\dfrac{2}{9} + \dfrac{4}{9} = \underline{}$

(36) $\dfrac{3}{7} + \dfrac{4}{7} = \underline{}$

(37) $\dfrac{3}{12} + \dfrac{3}{12} = \underline{}$

(38) $\dfrac{1}{4} + \dfrac{3}{4} = \underline{}$

(39) $\dfrac{2}{9} + \dfrac{8}{9} = \underline{}$

(40) $\dfrac{2}{3} + \dfrac{2}{3} = \underline{}$

(41) $\dfrac{3}{5} + \dfrac{5}{5} = \underline{}$

(42) $\dfrac{1}{8} + \dfrac{9}{8} = \underline{}$

(43) $\dfrac{3}{8} + \dfrac{2}{8} = \underline{}$

(44) $\dfrac{2}{6} + \dfrac{3}{6} = \underline{}$

(45) $\dfrac{2}{10} + \dfrac{3}{10} = \underline{}$

Adding fractions with like denominators:

(1) $\dfrac{5}{6} + \dfrac{1}{6} =$ ___

(2) $\dfrac{1}{10} + \dfrac{5}{10} =$ ___

(3) $\dfrac{1}{10} + \dfrac{10}{10} =$ ___

(4) $\dfrac{1}{10} + \dfrac{5}{10} =$ ___

(5) $\dfrac{2}{8} + \dfrac{5}{8} =$ ___

(6) $\dfrac{1}{7} + \dfrac{3}{7} =$ ___

(7) $\dfrac{7}{9} + \dfrac{5}{9} =$ ___

(8) $\dfrac{1}{12} + \dfrac{5}{12} =$ ___

(9) $\dfrac{2}{10} + \dfrac{7}{10} =$ ___

(10) $\dfrac{3}{5} + \dfrac{2}{5} =$ ___

(11) $\dfrac{3}{8} + \dfrac{5}{8} =$ ___

(12) $\dfrac{1}{10} + \dfrac{9}{10} =$ ___

(13) $\dfrac{2}{6} + \dfrac{5}{6} =$ ___

(14) $\dfrac{2}{12} + \dfrac{3}{12} =$ ___

(15) $\dfrac{2}{5} + \dfrac{4}{5} =$ ___

(16) $\dfrac{1}{3} + \dfrac{2}{3} =$ ___

(17) $\dfrac{1}{7} + \dfrac{3}{7} =$ ___

(18) $\dfrac{1}{9} + \dfrac{9}{9} =$ ___

(19) $\dfrac{4}{7} + \dfrac{4}{7} =$ ___

(20) $\dfrac{3}{9} + \dfrac{6}{9} =$ ___

(21) $\dfrac{2}{3} + \dfrac{4}{3} =$ ___

(22) $\dfrac{3}{4} + \dfrac{1}{4} =$ ___

(23) $\dfrac{2}{10} + \dfrac{6}{10} =$ ___

(24) $\dfrac{3}{8} + \dfrac{4}{8} =$ ___

(25) $\dfrac{5}{8} + \dfrac{2}{8} =$ ___

(26) $\dfrac{1}{6} + \dfrac{3}{6} =$ ___

(27) $\dfrac{1}{5} + \dfrac{4}{5} =$ ___

(28) $\dfrac{2}{4} + \dfrac{4}{4} =$ ___

(29) $\dfrac{3}{12} + \dfrac{5}{12} =$ ___

(30) $\dfrac{1}{4} + \dfrac{4}{4} =$ ___

(31) $\dfrac{1}{9} + \dfrac{4}{9} =$ ___

(32) $\dfrac{2}{7} + \dfrac{3}{7} =$ ___

(33) $\dfrac{1}{12} + \dfrac{7}{12} =$ ___

(34) $\dfrac{4}{10} + \dfrac{2}{10} =$ ___

(35) $\dfrac{1}{8} + \dfrac{3}{8} =$ ___

(36) $\dfrac{2}{7} + \dfrac{4}{7} =$ ___

(37) $\dfrac{3}{10} + \dfrac{4}{10} =$ ___

(38) $\dfrac{2}{6} + \dfrac{5}{6} =$ ___

(39) $\dfrac{2}{12} + \dfrac{5}{12} =$ ___

(40) $\dfrac{1}{6} + \dfrac{3}{6} =$ ___

(41) $\dfrac{1}{12} + \dfrac{3}{12} =$ ___

(42) $\dfrac{1}{3} + \dfrac{4}{3} =$ ___

(43) $\dfrac{2}{8} + \dfrac{4}{8} =$ ___

(44) $\dfrac{3}{4} + \dfrac{6}{4} =$ ___

(45) $\dfrac{2}{4} + \dfrac{3}{4} =$ ___

Adding fractions with like denominators:

(1) $\dfrac{3}{6} + \dfrac{3}{6} = $ ——

(2) $\dfrac{2}{12} + \dfrac{4}{12} = $ ——

(3) $\dfrac{1}{6} + \dfrac{6}{6} = $ ——

(4) $\dfrac{1}{7} + \dfrac{1}{7} = $ ——

(5) $\dfrac{1}{7} + \dfrac{5}{7} = $ ——

(6) $\dfrac{3}{5} + \dfrac{5}{5} = $ ——

(7) $\dfrac{2}{5} + \dfrac{2}{5} = $ ——

(8) $\dfrac{3}{6} + \dfrac{5}{6} = $ ——

(9) $\dfrac{1}{10} + \dfrac{8}{10} = $ ——

(10) $\dfrac{4}{9} + \dfrac{2}{9} = $ ——

(11) $\dfrac{2}{9} + \dfrac{3}{9} = $ ——

(12) $\dfrac{2}{8} + \dfrac{7}{8} = $ ——

(13) $\dfrac{1}{5} + \dfrac{4}{5} = $ ——

(14) $\dfrac{1}{10} + \dfrac{3}{10} = $ ——

(15) $\dfrac{1}{7} + \dfrac{5}{7} = $ ——

(16) $\dfrac{5}{7} + \dfrac{2}{7} = $ ——

(17) $\dfrac{3}{5} + \dfrac{6}{5} = $ ——

(18) $\dfrac{3}{12} + \dfrac{4}{12} = $ ——

(19) $\dfrac{3}{10} + \dfrac{3}{10} = $ ——

(20) $\dfrac{2}{8} + \dfrac{4}{8} = $ ——

(21) $\dfrac{2}{9} + \dfrac{6}{9} = $ ——

(22) $\dfrac{2}{12} + \dfrac{4}{12} = $ ——

(23) $\dfrac{1}{5} + \dfrac{3}{5} = $ ——

(24) $\dfrac{1}{8} + \dfrac{10}{8} = $ ——

(25) $\dfrac{1}{10} + \dfrac{3}{10} = $ ——

(26) $\dfrac{3}{7} + \dfrac{5}{7} = $ ——

(27) $\dfrac{2}{10} + \dfrac{4}{10} = $ ——

(28) $\dfrac{4}{6} + \dfrac{2}{6} = $ ——

(29) $\dfrac{2}{12} + \dfrac{4}{12} = $ ——

(30) $\dfrac{1}{9} + \dfrac{7}{9} = $ ——

(31) $\dfrac{3}{7} + \dfrac{1}{7} = $ ——

(32) $\dfrac{1}{4} + \dfrac{2}{4} = $ ——

(33) $\dfrac{2}{3} + \dfrac{5}{3} = $ ——

(34) $\dfrac{2}{9} + \dfrac{2}{9} = $ ——

(35) $\dfrac{3}{9} + \dfrac{4}{9} = $ ——

(36) $\dfrac{1}{3} + \dfrac{2}{3} = $ ——

(37) $\dfrac{1}{12} + \dfrac{3}{12} = $ ——

(38) $\dfrac{2}{10} + \dfrac{6}{10} = $ ——

(39) $\dfrac{2}{5} + \dfrac{1}{5} = $ ——

(40) $\dfrac{5}{10} + \dfrac{3}{10} = $ ——

(41) $\dfrac{1}{7} + \dfrac{6}{7} = $ ——

(42) $\dfrac{1}{6} + \dfrac{5}{6} = $ ——

(43) $\dfrac{2}{10} + \dfrac{2}{10} = $ ——

(44) $\dfrac{3}{8} + \dfrac{4}{8} = $ ——

(45) $\dfrac{3}{4} + \dfrac{1}{4} = $ ——

Adding fractions with like denominators:

(1) $\dfrac{1}{8} + \dfrac{2}{8} = $ ___

(2) $\dfrac{2}{10} + \dfrac{4}{10} = $ ___

(3) $\dfrac{2}{7} + \dfrac{5}{7} = $ ___

(4) $\dfrac{4}{8} + \dfrac{4}{8} = $ ___

(5) $\dfrac{1}{6} + \dfrac{2}{6} = $ ___

(6) $\dfrac{4}{9} + \dfrac{1}{9} = $ ___

(7) $\dfrac{3}{9} + \dfrac{4}{9} = $ ___

(8) $\dfrac{3}{10} + \dfrac{6}{10} = $ ___

(9) $\dfrac{3}{8} + \dfrac{1}{8} = $ ___

(10) $\dfrac{1}{3} + \dfrac{3}{3} = $ ___

(11) $\dfrac{2}{7} + \dfrac{4}{7} = $ ___

(12) $\dfrac{2}{10} + \dfrac{3}{10} = $ ___

(13) $\dfrac{2}{4} + \dfrac{3}{4} = $ ___

(14) $\dfrac{1}{8} + \dfrac{4}{8} = $ ___

(15) $\dfrac{5}{12} + \dfrac{2}{12} = $ ___

(16) $\dfrac{4}{5} + \dfrac{2}{5} = $ ___

(17) $\dfrac{2}{6} + \dfrac{4}{6} = $ ___

(18) $\dfrac{4}{10} + \dfrac{2}{10} = $ ___

(19) $\dfrac{3}{7} + \dfrac{5}{7} = $ ___

(20) $\dfrac{1}{12} + \dfrac{2}{12} = $ ___

(21) $\dfrac{1}{9} + \dfrac{4}{9} = $ ___

(22) $\dfrac{1}{6} + \dfrac{1}{6} = $ ___

(23) $\dfrac{3}{4} + \dfrac{3}{4} = $ ___

(24) $\dfrac{2}{8} + \dfrac{6}{8} = $ ___

(25) $\dfrac{2}{7} + \dfrac{5}{7} = $ ___

(26) $\dfrac{2}{7} + \dfrac{2}{7} = $ ___

(27) $\dfrac{1}{10} + \dfrac{3}{10} = $ ___

(28) $\dfrac{4}{12} + \dfrac{4}{12} = $ ___

(29) $\dfrac{3}{5} + \dfrac{3}{5} = $ ___

(30) $\dfrac{2}{6} + \dfrac{3}{6} = $ ___

(31) $\dfrac{3}{4} + \dfrac{4}{4} = $ ___

(32) $\dfrac{2}{9} + \dfrac{2}{9} = $ ___

(33) $\dfrac{1}{7} + \dfrac{3}{7} = $ ___

(34) $\dfrac{1}{10} + \dfrac{1}{10} = $ ___

(35) $\dfrac{1}{10} + \dfrac{2}{10} = $ ___

(36) $\dfrac{3}{10} + \dfrac{7}{10} = $ ___

(37) $\dfrac{2}{8} + \dfrac{2}{8} = $ ___

(38) $\dfrac{5}{7} + \dfrac{5}{7} = $ ___

(39) $\dfrac{2}{5} + \dfrac{3}{5} = $ ___

(40) $\dfrac{5}{9} + \dfrac{1}{9} = $ ___

(41) $\dfrac{3}{6} + \dfrac{5}{6} = $ ___

(42) $\dfrac{3}{12} + \dfrac{7}{12} = $ ___

(43) $\dfrac{1}{5} + \dfrac{2}{5} = $ ___

(44) $\dfrac{2}{8} + \dfrac{3}{8} = $ ___

(45) $\dfrac{4}{7} + \dfrac{3}{7} = $ ___

Adding fractions with like denominators:

(1) $\dfrac{3}{10} + \dfrac{5}{10} = $ —

(2) $\dfrac{1}{9} + \dfrac{2}{9} = $ —

(3) $\dfrac{5}{10} + \dfrac{6}{10} = $ —

(4) $\dfrac{2}{3} + \dfrac{3}{3} = $ —

(5) $\dfrac{4}{10} + \dfrac{5}{10} = $ —

(6) $\dfrac{1}{8} + \dfrac{6}{8} = $ —

(7) $\dfrac{4}{7} + \dfrac{1}{7} = $ —

(8) $\dfrac{2}{5} + \dfrac{3}{5} = $ —

(9) $\dfrac{2}{9} + \dfrac{5}{9} = $ —

(10) $\dfrac{1}{9} + \dfrac{2}{9} = $ —

(11) $\dfrac{1}{8} + \dfrac{2}{8} = $ —

(12) $\dfrac{2}{7} + \dfrac{5}{7} = $ —

(13) $\dfrac{3}{6} + \dfrac{2}{6} = $ —

(14) $\dfrac{4}{10} + \dfrac{5}{10} = $ —

(15) $\dfrac{4}{5} + \dfrac{1}{5} = $ —

(16) $\dfrac{2}{12} + \dfrac{2}{12} = $ —

(17) $\dfrac{2}{10} + \dfrac{4}{10} = $ —

(18) $\dfrac{3}{8} + \dfrac{4}{8} = $ —

(19) $\dfrac{4}{10} + \dfrac{4}{10} = $ —

(20) $\dfrac{1}{4} + \dfrac{2}{4} = $ —

(21) $\dfrac{4}{10} + \dfrac{7}{10} = $ —

(22) $\dfrac{2}{7} + \dfrac{3}{7} = $ —

(23) $\dfrac{3}{8} + \dfrac{5}{8} = $ —

(24) $\dfrac{1}{10} + \dfrac{3}{10} = $ —

(25) $\dfrac{4}{6} + \dfrac{2}{6} = $ —

(26) $\dfrac{2}{6} + \dfrac{3}{6} = $ —

(27) $\dfrac{5}{6} + \dfrac{2}{6} = $ —

(28) $\dfrac{1}{3} + \dfrac{3}{3} = $ —

(29) $\dfrac{1}{7} + \dfrac{2}{7} = $ —

(30) $\dfrac{4}{9} + \dfrac{5}{9} = $ —

(31) $\dfrac{3}{8} + \dfrac{4}{8} = $ —

(32) $\dfrac{4}{9} + \dfrac{5}{9} = $ —

(33) $\dfrac{1}{5} + \dfrac{4}{5} = $ —

(34) $\dfrac{2}{5} + \dfrac{4}{5} = $ —

(35) $\dfrac{3}{10} + \dfrac{5}{10} = $ —

(36) $\dfrac{3}{10} + \dfrac{4}{10} = $ —

(37) $\dfrac{1}{2} + \dfrac{4}{2} = $ —

(38) $\dfrac{2}{7} + \dfrac{4}{7} = $ —

(39) $\dfrac{2}{10} + \dfrac{3}{10} = $ —

(40) $\dfrac{4}{7} + \dfrac{5}{7} = $ —

(41) $\dfrac{1}{6} + \dfrac{2}{6} = $ —

(42) $\dfrac{3}{7} + \dfrac{2}{7} = $ —

(43) $\dfrac{2}{10} + \dfrac{3}{10} = $ —

(44) $\dfrac{4}{12} + \dfrac{5}{12} = $ —

(45) $\dfrac{1}{12} + \dfrac{3}{12} = $ —

Adding fractions with like denominators:

(1) $\dfrac{1}{7} + \dfrac{3}{7} = \underline{\quad}$

(2) $\dfrac{2}{10} + \dfrac{3}{10} = \underline{\quad}$

(3) $\dfrac{4}{10} + \dfrac{3}{10} = \underline{\quad}$

(4) $\dfrac{5}{6} + \dfrac{2}{6} = \underline{\quad}$

(5) $\dfrac{1}{10} + \dfrac{2}{10} = \underline{\quad}$

(6) $\dfrac{5}{8} + \dfrac{1}{8} = \underline{\quad}$

(7) $\dfrac{3}{5} + \dfrac{4}{5} = \underline{\quad}$

(8) $\dfrac{3}{7} + \dfrac{4}{7} = \underline{\quad}$

(9) $\dfrac{2}{12} + \dfrac{5}{12} = \underline{\quad}$

(10) $\dfrac{1}{8} + \dfrac{3}{8} = \underline{\quad}$

(11) $\dfrac{2}{3} + \dfrac{3}{3} = \underline{\quad}$

(12) $\dfrac{2}{4} + \dfrac{1}{4} = \underline{\quad}$

(13) $\dfrac{4}{12} + \dfrac{5}{12} = \underline{\quad}$

(14) $\dfrac{1}{12} + \dfrac{2}{12} = \underline{\quad}$

(15) $\dfrac{1}{10} + \dfrac{2}{10} = \underline{\quad}$

(16) $\dfrac{2}{6} + \dfrac{3}{6} = \underline{\quad}$

(17) $\dfrac{4}{8} + \dfrac{5}{8} = \underline{\quad}$

(18) $\dfrac{3}{6} + \dfrac{2}{6} = \underline{\quad}$

(19) $\dfrac{1}{5} + \dfrac{3}{5} = \underline{\quad}$

(20) $\dfrac{2}{4} + \dfrac{3}{4} = \underline{\quad}$

(21) $\dfrac{2}{3} + \dfrac{1}{3} = \underline{\quad}$

(22) $\dfrac{3}{10} + \dfrac{5}{10} = \underline{\quad}$

(23) $\dfrac{1}{5} + \dfrac{2}{5} = \underline{\quad}$

(24) $\dfrac{1}{4} + \dfrac{3}{4} = \underline{\quad}$

(25) $\dfrac{2}{8} + \dfrac{3}{8} = \underline{\quad}$

(26) $\dfrac{4}{10} + \dfrac{5}{10} = \underline{\quad}$

(27) $\dfrac{4}{8} + \dfrac{6}{8} = \underline{\quad}$

(28) $\dfrac{1}{6} + \dfrac{2}{6} = \underline{\quad}$

(29) $\dfrac{2}{9} + \dfrac{3}{9} = \underline{\quad}$

(30) $\dfrac{5}{9} + \dfrac{3}{9} = \underline{\quad}$

(31) $\dfrac{4}{9} + \dfrac{5}{9} = \underline{\quad}$

(32) $\dfrac{1}{2} + \dfrac{2}{2} = \underline{\quad}$

(33) $\dfrac{3}{5} + \dfrac{2}{5} = \underline{\quad}$

(34) $\dfrac{2}{10} + \dfrac{3}{10} = \underline{\quad}$

(35) $\dfrac{3}{6} + \dfrac{4}{6} = \underline{\quad}$

(36) $\dfrac{1}{6} + \dfrac{4}{6} = \underline{\quad}$

(37) $\dfrac{1}{12} + \dfrac{2}{12} = \underline{\quad}$

(38) $\dfrac{2}{12} + \dfrac{3}{12} = \underline{\quad}$

(39) $\dfrac{2}{10} + \dfrac{6}{10} = \underline{\quad}$

(40) $\dfrac{3}{9} + \dfrac{3}{9} = \underline{\quad}$

(41) $\dfrac{1}{10} + \dfrac{3}{10} = \underline{\quad}$

(42) $\dfrac{2}{12} + \dfrac{1}{12} = \underline{\quad}$

(43) $\dfrac{2}{4} + \dfrac{5}{4} = \underline{\quad}$

(44) $\dfrac{4}{6} + \dfrac{4}{6} = \underline{\quad}$

(45) $\dfrac{4}{6} + \dfrac{2}{6} = \underline{\quad}$

Adding fractions with like denominators:

(1) $\dfrac{1}{12} + \dfrac{2}{12} = $ ———

(2) $\dfrac{2}{7} + \dfrac{2}{7} = $ ———

(3) $\dfrac{1}{8} + \dfrac{7}{8} = $ ———

(4) $\dfrac{4}{8} + \dfrac{5}{8} = $ ———

(5) $\dfrac{1}{3} + \dfrac{2}{3} = $ ———

(6) $\dfrac{5}{7} + \dfrac{4}{7} = $ ———

(7) $\dfrac{2}{9} + \dfrac{3}{9} = $ ———

(8) $\dfrac{3}{9} + \dfrac{5}{9} = $ ———

(9) $\dfrac{4}{12} + \dfrac{6}{12} = $ ———

(10) $\dfrac{1}{10} + \dfrac{2}{10} = $ ———

(11) $\dfrac{2}{5} + \dfrac{4}{5} = $ ———

(12) $\dfrac{3}{4} + \dfrac{2}{4} = $ ———

(13) $\dfrac{5}{12} + \dfrac{4}{12} = $ ———

(14) $\dfrac{1}{10} + \dfrac{3}{10} = $ ———

(15) $\dfrac{3}{9} + \dfrac{6}{9} = $ ———

(16) $\dfrac{3}{7} + \dfrac{4}{7} = $ ———

(17) $\dfrac{4}{7} + \dfrac{4}{7} = $ ———

(18) $\dfrac{5}{10} + \dfrac{3}{10} = $ ———

(19) $\dfrac{2}{5} + \dfrac{5}{5} = $ ———

(20) $\dfrac{2}{8} + \dfrac{4}{8} = $ ———

(21) $\dfrac{1}{3} + \dfrac{5}{3} = $ ———

(22) $\dfrac{1}{4} + \dfrac{3}{4} = $ ———

(23) $\dfrac{1}{6} + \dfrac{3}{6} = $ ———

(24) $\dfrac{2}{8} + \dfrac{4}{8} = $ ———

(25) $\dfrac{4}{6} + \dfrac{3}{6} = $ ———

(26) $\dfrac{3}{5} + \dfrac{5}{5} = $ ———

(27) $\dfrac{1}{9} + \dfrac{2}{9} = $ ———

(28) $\dfrac{2}{7} + \dfrac{4}{7} = $ ———

(29) $\dfrac{2}{10} + \dfrac{5}{10} = $ ———

(30) $\dfrac{5}{10} + \dfrac{2}{10} = $ ———

(31) $\dfrac{1}{9} + \dfrac{1}{9} = $ ———

(32) $\dfrac{1}{4} + \dfrac{3}{4} = $ ———

(33) $\dfrac{4}{10} + \dfrac{6}{10} = $ ———

(34) $\dfrac{3}{10} + \dfrac{4}{10} = $ ———

(35) $\dfrac{4}{9} + \dfrac{4}{9} = $ ———

(36) $\dfrac{1}{2} + \dfrac{1}{2} = $ ———

(37) $\dfrac{2}{12} + \dfrac{3}{12} = $ ———

(38) $\dfrac{2}{6} + \dfrac{4}{6} = $ ———

(39) $\dfrac{4}{6} + \dfrac{2}{6} = $ ———

(40) $\dfrac{1}{7} + \dfrac{2}{7} = $ ———

(41) $\dfrac{1}{9} + \dfrac{3}{9} = $ ———

(42) $\dfrac{2}{8} + \dfrac{1}{8} = $ ———

(43) $\dfrac{4}{12} + \dfrac{5}{12} = $ ———

(44) $\dfrac{3}{7} + \dfrac{5}{7} = $ ———

(45) $\dfrac{3}{4} + \dfrac{1}{4} = $ ———

Adding fractions with like denominators:

(1) $\dfrac{2}{6} + \dfrac{4}{6} = $ —

(2) $\dfrac{2}{4} + \dfrac{4}{4} = $ —

(3) $\dfrac{3}{10} + \dfrac{2}{10} = $ —

(4) $\dfrac{1}{3} + \dfrac{2}{3} = $ —

(5) $\dfrac{1}{12} + \dfrac{3}{12} = $ —

(6) $\dfrac{5}{9} + \dfrac{4}{9} = $ —

(7) $\dfrac{3}{8} + \dfrac{5}{8} = $ —

(8) $\dfrac{4}{8} + \dfrac{4}{8} = $ —

(9) $\dfrac{3}{7} + \dfrac{1}{7} = $ —

(10) $\dfrac{2}{4} + \dfrac{3}{4} = $ —

(11) $\dfrac{2}{10} + \dfrac{4}{10} = $ —

(12) $\dfrac{4}{12} + \dfrac{3}{12} = $ —

(13) $\dfrac{1}{5} + \dfrac{1}{5} = $ —

(14) $\dfrac{1}{10} + \dfrac{3}{10} = $ —

(15) $\dfrac{5}{8} + \dfrac{4}{8} = $ —

(16) $\dfrac{4}{9} + \dfrac{4}{9} = $ —

(17) $\dfrac{3}{8} + \dfrac{4}{8} = $ —

(18) $\dfrac{2}{5} + \dfrac{1}{5} = $ —

(19) $\dfrac{2}{10} + \dfrac{4}{10} = $ —

(20) $\dfrac{2}{5} + \dfrac{5}{5} = $ —

(21) $\dfrac{5}{6} + \dfrac{2}{6} = $ —

(22) $\dfrac{1}{12} + \dfrac{3}{12} = $ —

(23) $\dfrac{1}{7} + \dfrac{3}{7} = $ —

(24) $\dfrac{1}{3} + \dfrac{1}{3} = $ —

(25) $\dfrac{3}{7} + \dfrac{3}{7} = $ —

(26) $\dfrac{3}{10} + \dfrac{4}{10} = $ —

(27) $\dfrac{5}{7} + \dfrac{3}{7} = $ —

(28) $\dfrac{2}{8} + \dfrac{4}{8} = $ —

(29) $\dfrac{2}{9} + \dfrac{4}{9} = $ —

(30) $\dfrac{3}{9} + \dfrac{1}{9} = $ —

(31) $\dfrac{1}{6} + \dfrac{3}{6} = $ —

(32) $\dfrac{2}{3} + \dfrac{3}{3} = $ —

(33) $\dfrac{2}{7} + \dfrac{1}{7} = $ —

(34) $\dfrac{4}{12} + \dfrac{5}{12} = $ —

(35) $\dfrac{4}{9} + \dfrac{2}{9} = $ —

(36) $\dfrac{4}{8} + \dfrac{3}{8} = $ —

(37) $\dfrac{2}{7} + \dfrac{2}{7} = $ —

(38) $\dfrac{1}{10} + \dfrac{1}{10} = $ —

(39) $\dfrac{3}{12} + \dfrac{2}{12} = $ —

(40) $\dfrac{4}{10} + \dfrac{5}{10} = $ —

(41) $\dfrac{3}{7} + \dfrac{5}{7} = $ —

(42) $\dfrac{2}{6} + \dfrac{1}{6} = $ —

(43) $\dfrac{1}{12} + \dfrac{1}{12} = $ —

(44) $\dfrac{2}{6} + \dfrac{4}{6} = $ —

(45) $\dfrac{4}{10} + \dfrac{2}{10} = $ —

Adding fractions with like denominators:

(1) $\dfrac{3}{9} + \dfrac{5}{9} = \underline{\quad}$

(2) $\dfrac{1}{5} + \dfrac{2}{5} = \underline{\quad}$

(3) $\dfrac{1}{8} + \dfrac{1}{8} = \underline{\quad}$

(4) $\dfrac{2}{5} + \dfrac{3}{5} = \underline{\quad}$

(5) $\dfrac{4}{8} + \dfrac{4}{8} = \underline{\quad}$

(6) $\dfrac{5}{12} + \dfrac{4}{12} = \underline{\quad}$

(7) $\dfrac{1}{8} + \dfrac{2}{8} = \underline{\quad}$

(8) $\dfrac{2}{4} + \dfrac{5}{4} = \underline{\quad}$

(9) $\dfrac{4}{9} + \dfrac{1}{9} = \underline{\quad}$

(10) $\dfrac{4}{7} + \dfrac{4}{7} = \underline{\quad}$

(11) $\dfrac{1}{12} + \dfrac{3}{12} = \underline{\quad}$

(12) $\dfrac{2}{4} + \dfrac{1}{4} = \underline{\quad}$

(13) $\dfrac{3}{12} + \dfrac{3}{12} = \underline{\quad}$

(14) $\dfrac{5}{10} + \dfrac{2}{10} = \underline{\quad}$

(15) $\dfrac{3}{6} + \dfrac{2}{6} = \underline{\quad}$

(16) $\dfrac{2}{3} + \dfrac{4}{3} = \underline{\quad}$

(17) $\dfrac{3}{8} + \dfrac{3}{8} = \underline{\quad}$

(18) $\dfrac{4}{7} + \dfrac{3}{7} = \underline{\quad}$

(19) $\dfrac{1}{4} + \dfrac{4}{4} = \underline{\quad}$

(20) $\dfrac{1}{7} + \dfrac{1}{7} = \underline{\quad}$

(21) $\dfrac{1}{10} + \dfrac{1}{10} = \underline{\quad}$

(22) $\dfrac{2}{5} + \dfrac{2}{5} = \underline{\quad}$

(23) $\dfrac{4}{5} + \dfrac{5}{5} = \underline{\quad}$

(24) $\dfrac{3}{8} + \dfrac{2}{8} = \underline{\quad}$

(25) $\dfrac{3}{7} + \dfrac{3}{7} = \underline{\quad}$

(26) $\dfrac{2}{10} + \dfrac{4}{10} = \underline{\quad}$

(27) $\dfrac{4}{5} + \dfrac{2}{5} = \underline{\quad}$

(28) $\dfrac{4}{12} + \dfrac{4}{12} = \underline{\quad}$

(29) $\dfrac{1}{6} + \dfrac{1}{6} = \underline{\quad}$

(30) $\dfrac{1}{7} + \dfrac{1}{7} = \underline{\quad}$

(31) $\dfrac{1}{10} + \dfrac{1}{10} = \underline{\quad}$

(32) $\dfrac{3}{9} + \dfrac{3}{9} = \underline{\quad}$

(33) $\dfrac{2}{9} + \dfrac{1}{9} = \underline{\quad}$

(34) $\dfrac{5}{9} + \dfrac{5}{9} = \underline{\quad}$

(35) $\dfrac{2}{7} + \dfrac{3}{7} = \underline{\quad}$

(36) $\dfrac{4}{12} + \dfrac{3}{12} = \underline{\quad}$

(37) $\dfrac{2}{4} + \dfrac{2}{4} = \underline{\quad}$

(38) $\dfrac{1}{10} + \dfrac{1}{10} = \underline{\quad}$

(39) $\dfrac{5}{12} + \dfrac{4}{12} = \underline{\quad}$

(40) $\dfrac{3}{8} + \dfrac{3}{8} = \underline{\quad}$

(41) $\dfrac{4}{6} + \dfrac{5}{6} = \underline{\quad}$

(42) $\dfrac{3}{12} + \dfrac{2}{12} = \underline{\quad}$

(43) $\dfrac{4}{12} + \dfrac{4}{12} = \underline{\quad}$

(44) $\dfrac{2}{12} + \dfrac{4}{12} = \underline{\quad}$

(45) $\dfrac{2}{12} + \dfrac{1}{12} = \underline{\quad}$

Adding fractions with like denominators:

(1) $\dfrac{1}{7} + \dfrac{1}{7} = \underline{}$

(2) $\dfrac{1}{8} + \dfrac{1}{8} = \underline{}$

(3) $\dfrac{1}{9} + \dfrac{1}{9} = \underline{}$

(4) $\dfrac{5}{12} + \dfrac{5}{12} = \underline{}$

(5) $\dfrac{3}{6} + \dfrac{4}{6} = \underline{}$

(6) $\dfrac{3}{5} + \dfrac{2}{5} = \underline{}$

(7) $\dfrac{2}{9} + \dfrac{2}{9} = \underline{}$

(8) $\dfrac{2}{8} + \dfrac{5}{8} = \underline{}$

(9) $\dfrac{2}{10} + \dfrac{1}{10} = \underline{}$

(10) $\dfrac{3}{10} + \dfrac{3}{10} = \underline{}$

(11) $\dfrac{1}{4} + \dfrac{1}{4} = \underline{}$

(12) $\dfrac{4}{4} + \dfrac{3}{4} = \underline{}$

(13) $\dfrac{4}{13} + \dfrac{4}{13} = \underline{}$

(14) $\dfrac{4}{10} + \dfrac{4}{10} = \underline{}$

(15) $\dfrac{1}{12} + \dfrac{1}{12} = \underline{}$

(16) $\dfrac{1}{8} + \dfrac{1}{8} = \underline{}$

(17) $\dfrac{2}{9} + \dfrac{4}{9} = \underline{}$

(18) $\dfrac{3}{3} + \dfrac{2}{3} = \underline{}$

(19) $\dfrac{5}{14} + \dfrac{5}{14} = \underline{}$

(20) $\dfrac{1}{3} + \dfrac{1}{3} = \underline{}$

(21) $\dfrac{4}{8} + \dfrac{2}{8} = \underline{}$

(22) $\dfrac{2}{7} + \dfrac{2}{7} = \underline{}$

(23) $\dfrac{3}{10} + \dfrac{3}{10} = \underline{}$

(24) $\dfrac{5}{10} + \dfrac{4}{10} = \underline{}$

(25) $\dfrac{3}{9} + \dfrac{3}{9} = \underline{}$

(26) $\dfrac{2}{5} + \dfrac{2}{5} = \underline{}$

(27) $\dfrac{2}{12} + \dfrac{1}{12} = \underline{}$

(28) $\dfrac{4}{10} + \dfrac{4}{10} = \underline{}$

(29) $\dfrac{1}{9} + \dfrac{1}{9} = \underline{}$

(30) $\dfrac{1}{6} + \dfrac{1}{6} = \underline{}$

(31) $\dfrac{1}{13} + \dfrac{1}{13} = \underline{}$

(32) $\dfrac{4}{10} + \dfrac{4}{10} = \underline{}$

(33) $\dfrac{3}{7} + \dfrac{2}{7} = \underline{}$

(34) $\dfrac{5}{12} + \dfrac{5}{12} = \underline{}$

(35) $\dfrac{2}{6} + \dfrac{5}{6} = \underline{}$

(36) $\dfrac{4}{6} + \dfrac{3}{6} = \underline{}$

(37) $\dfrac{2}{6} + \dfrac{2}{6} = \underline{}$

(38) $\dfrac{1}{2} + \dfrac{1}{2} = \underline{}$

(39) $\dfrac{2}{3} + \dfrac{1}{3} = \underline{}$

(40) $\dfrac{3}{12} + \dfrac{3}{12} = \underline{}$

(41) $\dfrac{3}{5} + \dfrac{3}{5} = \underline{}$

(42) $\dfrac{1}{12} + \dfrac{1}{12} = \underline{}$

(43) $\dfrac{4}{8} + \dfrac{4}{8} = \underline{}$

(44) $\dfrac{2}{3} + \dfrac{5}{3} = \underline{}$

(45) $\dfrac{3}{4} + \dfrac{2}{4} = \underline{}$

Adding fractions with like denominators:

(1) $\dfrac{1}{5} + \dfrac{1}{5} = \underline{\quad}$

(2) $\dfrac{1}{12} + \dfrac{1}{12} = \underline{\quad}$

(3) $\dfrac{5}{8} + \dfrac{5}{8} = \underline{\quad}$

(4) $\dfrac{5}{10} + \dfrac{5}{10} = \underline{\quad}$

(5) $\dfrac{4}{9} + \dfrac{4}{9} = \underline{\quad}$

(6) $\dfrac{1}{7} + \dfrac{3}{7} = \underline{\quad}$

(7) $\dfrac{2}{12} + \dfrac{2}{12} = \underline{\quad}$

(8) $\dfrac{2}{10} + \dfrac{2}{10} = \underline{\quad}$

(9) $\dfrac{3}{8} + \dfrac{5}{8} = \underline{\quad}$

(10) $\dfrac{3}{13} + \dfrac{3}{13} = \underline{\quad}$

(11) $\dfrac{1}{10} + \dfrac{1}{10} = \underline{\quad}$

(12) $\dfrac{2}{12} + \dfrac{3}{12} = \underline{\quad}$

(13) $\dfrac{4}{9} + \dfrac{4}{9} = \underline{\quad}$

(14) $\dfrac{3}{7} + \dfrac{3}{7} = \underline{\quad}$

(15) $\dfrac{1}{6} + \dfrac{2}{6} = \underline{\quad}$

(16) $\dfrac{1}{12} + \dfrac{1}{12} = \underline{\quad}$

(17) $\dfrac{2}{8} + \dfrac{2}{8} = \underline{\quad}$

(18) $\dfrac{2}{9} + \dfrac{7}{9} = \underline{\quad}$

(19) $\dfrac{5}{8} + \dfrac{5}{8} = \underline{\quad}$

(20) $\dfrac{1}{6} + \dfrac{1}{6} = \underline{\quad}$

(21) $\dfrac{1}{12} + \dfrac{1}{12} = \underline{\quad}$

(22) $\dfrac{2}{8} + \dfrac{2}{8} = \underline{\quad}$

(23) $\dfrac{4}{12} + \dfrac{4}{12} = \underline{\quad}$

(24) $\dfrac{1}{8} + \dfrac{7}{8} = \underline{\quad}$

(25) $\dfrac{3}{12} + \dfrac{3}{12} = \underline{\quad}$

(26) $\dfrac{2}{7} + \dfrac{5}{7} = \underline{\quad}$

(27) $\dfrac{3}{6} + \dfrac{2}{6} = \underline{\quad}$

(28) $\dfrac{4}{15} + \dfrac{4}{15} = \underline{\quad}$

(29) $\dfrac{1}{5} + \dfrac{1}{5} = \underline{\quad}$

(30) $\dfrac{2}{3} + \dfrac{3}{3} = \underline{\quad}$

(31) $\dfrac{1}{12} + \dfrac{1}{12} = \underline{\quad}$

(32) $\dfrac{3}{10} + \dfrac{4}{10} = \underline{\quad}$

(33) $\dfrac{1}{4} + \dfrac{3}{4} = \underline{\quad}$

(34) $\dfrac{5}{13} + \dfrac{5}{13} = \underline{\quad}$

(35) $\dfrac{2}{4} + \dfrac{2}{4} = \underline{\quad}$

(36) $\dfrac{1}{9} + \dfrac{8}{9} = \underline{\quad}$

(37) $\dfrac{2}{7} + \dfrac{2}{7} = \underline{\quad}$

(38) $\dfrac{1}{10} + \dfrac{2}{10} = \underline{\quad}$

(39) $\dfrac{4}{10} + \dfrac{1}{10} = \underline{\quad}$

(40) $\dfrac{3}{10} + \dfrac{3}{10} = \underline{\quad}$

(41) $\dfrac{4}{7} + \dfrac{5}{7} = \underline{\quad}$

(42) $\dfrac{1}{2} + \dfrac{4}{2} = \underline{\quad}$

(43) $\dfrac{4}{13} + \dfrac{4}{13} = \underline{\quad}$

(44) $\dfrac{2}{12} + \dfrac{3}{12} = \underline{\quad}$

(45) $\dfrac{4}{8} + \dfrac{4}{8} = \underline{\quad}$

Adding fractions with like denominators:

(1) $\dfrac{1}{9} + \dfrac{1}{9} = \underline{}$

(2) $\dfrac{1}{4} + \dfrac{1}{4} = \underline{}$

(3) $\dfrac{1}{5} + \dfrac{4}{5} = \underline{}$

(4) $\dfrac{5}{12} + \dfrac{5}{12} = \underline{}$

(5) $\dfrac{3}{8} + \dfrac{4}{8} = \underline{}$

(6) $\dfrac{1}{10} + \dfrac{4}{10} = \underline{}$

(7) $\dfrac{2}{6} + \dfrac{2}{6} = \underline{}$

(8) $\dfrac{2}{5} + \dfrac{4}{5} = \underline{}$

(9) $\dfrac{2}{10} + \dfrac{3}{10} = \underline{}$

(10) $\dfrac{3}{9} + \dfrac{3}{9} = \underline{}$

(11) $\dfrac{1}{3} + \dfrac{1}{3} = \underline{}$

(12) $\dfrac{1}{12} + \dfrac{2}{12} = \underline{}$

(13) $\dfrac{4}{12} + \dfrac{4}{12} = \underline{}$

(14) $\dfrac{3}{9} + \dfrac{4}{9} = \underline{}$

(15) $\dfrac{1}{8} + \dfrac{5}{8} = \underline{}$

(16) $\dfrac{1}{8} + \dfrac{1}{8} = \underline{}$

(17) $\dfrac{2}{6} + \dfrac{3}{6} = \underline{}$

(18) $\dfrac{3}{5} + \dfrac{1}{5} = \underline{}$

(19) $\dfrac{5}{10} + \dfrac{5}{10} = \underline{}$

(20) $\dfrac{1}{7} + \dfrac{2}{7} = \underline{}$

(21) $\dfrac{3}{9} + \dfrac{5}{9} = \underline{}$

(22) $\dfrac{2}{5} + \dfrac{2}{5} = \underline{}$

(23) $\dfrac{4}{8} + \dfrac{5}{8} = \underline{}$

(24) $\dfrac{1}{6} + \dfrac{5}{6} = \underline{}$

(25) $\dfrac{3}{7} + \dfrac{3}{7} = \underline{}$

(26) $\dfrac{2}{12} + \dfrac{3}{12} = \underline{}$

(27) $\dfrac{2}{6} + \dfrac{5}{6} = \underline{}$

(28) $\dfrac{4}{10} + \dfrac{4}{10} = \underline{}$

(29) $\dfrac{1}{8} + \dfrac{2}{8} = \underline{}$

(30) $\dfrac{1}{12} + \dfrac{5}{12} = \underline{}$

(31) $\dfrac{1}{9} + \dfrac{1}{9} = \underline{}$

(32) $\dfrac{2}{7} + \dfrac{3}{7} = \underline{}$

(33) $\dfrac{1}{4} + \dfrac{5}{4} = \underline{}$

(34) $\dfrac{5}{12} + \dfrac{5}{12} = \underline{}$

(35) $\dfrac{4}{6} + \dfrac{2}{6} = \underline{}$

(36) $\dfrac{1}{7} + \dfrac{4}{7} = \underline{}$

(37) $\dfrac{2}{8} + \dfrac{2}{8} = \underline{}$

(38) $\dfrac{1}{3} + \dfrac{3}{3} = \underline{}$

(39) $\dfrac{4}{6} + \dfrac{2}{6} = \underline{}$

(40) $\dfrac{3}{12} + \dfrac{3}{12} = \underline{}$

(41) $\dfrac{3}{8} + \dfrac{4}{8} = \underline{}$

(42) $\dfrac{3}{12} + \dfrac{3}{12} = \underline{}$

(43) $\dfrac{4}{6} + \dfrac{4}{6} = \underline{}$

(44) $\dfrac{2}{5} + \dfrac{4}{5} = \underline{}$

(45) $\dfrac{1}{3} + \dfrac{3}{3} = \underline{}$

Adding fractions with like denominators:

(1) $\dfrac{1}{7} + \dfrac{1}{7} = \underline{\quad}$

(2) $\dfrac{1}{2} + \dfrac{4}{2} = \underline{\quad}$

(3) $\dfrac{1}{5} + \dfrac{3}{5} = \underline{\quad}$

(4) $\dfrac{5}{10} + \dfrac{5}{10} = \underline{\quad}$

(5) $\dfrac{4}{7} + \dfrac{5}{7} = \underline{\quad}$

(6) $\dfrac{2}{5} + \dfrac{5}{5} = \underline{\quad}$

(7) $\dfrac{2}{5} + \dfrac{2}{5} = \underline{\quad}$

(8) $\dfrac{2}{8} + \dfrac{3}{8} = \underline{\quad}$

(9) $\dfrac{1}{10} + \dfrac{5}{10} = \underline{\quad}$

(10) $\dfrac{3}{8} + \dfrac{3}{8} = \underline{\quad}$

(11) $\dfrac{1}{7} + \dfrac{3}{7} = \underline{\quad}$

(12) $\dfrac{1}{9} + \dfrac{2}{9} = \underline{\quad}$

(13) $\dfrac{4}{12} + \dfrac{4}{12} = \underline{\quad}$

(14) $\dfrac{5}{6} + \dfrac{2}{6} = \underline{\quad}$

(15) $\dfrac{2}{7} + \dfrac{2}{7} = \underline{\quad}$

(16) $\dfrac{1}{12} + \dfrac{1}{12} = \underline{\quad}$

(17) $\dfrac{3}{5} + \dfrac{4}{5} = \underline{\quad}$

(18) $\dfrac{1}{12} + \dfrac{1}{12} = \underline{\quad}$

(19) $\dfrac{2}{7} + \dfrac{2}{7} = \underline{\quad}$

(20) $\dfrac{1}{8} + \dfrac{3}{8} = \underline{\quad}$

(21) $\dfrac{1}{7} + \dfrac{5}{7} = \underline{\quad}$

(22) $\dfrac{3}{5} + \dfrac{3}{5} = \underline{\quad}$

(23) $\dfrac{4}{12} + \dfrac{5}{12} = \underline{\quad}$

(24) $\dfrac{2}{8} + \dfrac{2}{8} = \underline{\quad}$

(25) $\dfrac{4}{9} + \dfrac{4}{9} = \underline{\quad}$

(26) $\dfrac{2}{6} + \dfrac{3}{6} = \underline{\quad}$

(27) $\dfrac{1}{3} + \dfrac{5}{3} = \underline{\quad}$

(28) $\dfrac{1}{6} + \dfrac{1}{6} = \underline{\quad}$

(29) $\dfrac{1}{5} + \dfrac{3}{5} = \underline{\quad}$

(30) $\dfrac{2}{3} + \dfrac{5}{3} = \underline{\quad}$

(31) $\dfrac{5}{12} + \dfrac{5}{12} = \underline{\quad}$

(32) $\dfrac{3}{8} + \dfrac{5}{8} = \underline{\quad}$

(33) $\dfrac{1}{6} + \dfrac{3}{6} = \underline{\quad}$

(34) $\dfrac{2}{3} + \dfrac{2}{3} = \underline{\quad}$

(35) $\dfrac{2}{8} + \dfrac{3}{8} = \underline{\quad}$

(36) $\dfrac{3}{10} + \dfrac{1}{10} = \underline{\quad}$

(37) $\dfrac{3}{4} + \dfrac{3}{4} = \underline{\quad}$

(38) $\dfrac{1}{6} + \dfrac{2}{6} = \underline{\quad}$

(39) $\dfrac{1}{8} + \dfrac{4}{8} = \underline{\quad}$

(40) $\dfrac{4}{5} + \dfrac{4}{5} = \underline{\quad}$

(41) $\dfrac{4}{9} + \dfrac{5}{9} = \underline{\quad}$

(42) $\dfrac{2}{4} + \dfrac{2}{4} = \underline{\quad}$

(43) $\dfrac{1}{12} + \dfrac{1}{12} = \underline{\quad}$

(44) $\dfrac{2}{10} + \dfrac{3}{10} = \underline{\quad}$

(45) $\dfrac{1}{4} + \dfrac{4}{4} = \underline{\quad}$

Adding fractions with like denominators:

(1) $\dfrac{5}{8} + \dfrac{5}{8} = $ —

(2) $\dfrac{1}{8} + \dfrac{2}{8} = $ —

(3) $\dfrac{1}{12} + \dfrac{4}{12} = $ —

(4) $\dfrac{2}{9} + \dfrac{2}{9} = $ —

(5) $\dfrac{3}{9} + \dfrac{3}{9} = $ —

(6) $\dfrac{3}{8} + \dfrac{3}{8} = $ —

(7) $\dfrac{3}{10} + \dfrac{3}{10} = $ —

(8) $\dfrac{2}{4} + \dfrac{5}{4} = $ —

(9) $\dfrac{1}{9} + \dfrac{5}{9} = $ —

(10) $\dfrac{4}{7} + \dfrac{4}{7} = $ —

(11) $\dfrac{1}{12} + \dfrac{2}{12} = $ —

(12) $\dfrac{3}{4} + \dfrac{2}{4} = $ —

(13) $\dfrac{1}{9} + \dfrac{1}{9} = $ —

(14) $\dfrac{4}{8} + \dfrac{5}{8} = $ —

(15) $\dfrac{2}{9} + \dfrac{4}{9} = $ —

(16) $\dfrac{5}{15} + \dfrac{5}{15} = $ —

(17) $\dfrac{2}{9} + \dfrac{3}{9} = $ —

(18) $\dfrac{1}{5} + \dfrac{5}{5} = $ —

(19) $\dfrac{2}{4} + \dfrac{2}{4} = $ —

(20) $\dfrac{1}{10} + \dfrac{2}{10} = $ —

(21) $\dfrac{1}{12} + \dfrac{5}{12} = $ —

(22) $\dfrac{3}{7} + \dfrac{3}{7} = $ —

(23) $\dfrac{5}{8} + \dfrac{4}{8} = $ —

(24) $\dfrac{2}{12} + \dfrac{2}{12} = $ —

(25) $\dfrac{4}{10} + \dfrac{4}{10} = $ —

(26) $\dfrac{3}{7} + \dfrac{4}{7} = $ —

(27) $\dfrac{2}{12} + \dfrac{1}{12} = $ —

(28) $\dfrac{1}{12} + \dfrac{1}{12} = $ —

(29) $\dfrac{2}{5} + \dfrac{5}{5} = $ —

(30) $\dfrac{1}{3} + \dfrac{4}{3} = $ —

(31) $\dfrac{5}{13} + \dfrac{5}{13} = $ —

(32) $\dfrac{1}{4} + \dfrac{3}{4} = $ —

(33) $\dfrac{2}{3} + \dfrac{4}{3} = $ —

(34) $\dfrac{2}{6} + \dfrac{2}{6} = $ —

(35) $\dfrac{4}{6} + \dfrac{3}{6} = $ —

(36) $\dfrac{1}{7} + \dfrac{2}{7} = $ —

(37) $\dfrac{3}{9} + \dfrac{3}{9} = $ —

(38) $\dfrac{2}{7} + \dfrac{4}{7} = $ —

(39) $\dfrac{3}{5} + \dfrac{2}{5} = $ —

(40) $\dfrac{4}{12} + \dfrac{4}{12} = $ —

(41) $\dfrac{1}{9} + \dfrac{1}{9} = $ —

(42) $\dfrac{1}{10} + \dfrac{3}{10} = $ —

(43) $\dfrac{1}{5} + \dfrac{1}{5} = $ —

(44) $\dfrac{3}{10} + \dfrac{4}{10} = $ —

(45) $\dfrac{1}{6} + \dfrac{4}{6} = $ —

Adding fractions with like denominators:

(1) $\dfrac{5}{7} + \dfrac{5}{7} = \underline{\quad}$

(2) $\dfrac{2}{12} + \dfrac{3}{12} = \underline{\quad}$

(3) $\dfrac{2}{6} + \dfrac{4}{6} = \underline{\quad}$

(4) $\dfrac{2}{8} + \dfrac{2}{8} = \underline{\quad}$

(5) $\dfrac{1}{7} + \dfrac{2}{7} = \underline{\quad}$

(6) $\dfrac{1}{4} + \dfrac{2}{4} = \underline{\quad}$

(7) $\dfrac{3}{12} + \dfrac{3}{12} = \underline{\quad}$

(8) $\dfrac{4}{8} + \dfrac{5}{8} = \underline{\quad}$

(9) $\dfrac{2}{9} + \dfrac{3}{9} = \underline{\quad}$

(10) $\dfrac{4}{6} + \dfrac{4}{6} = \underline{\quad}$

(11) $\dfrac{2}{6} + \dfrac{4}{6} = \underline{\quad}$

(12) $\dfrac{3}{7} + \dfrac{1}{7} = \underline{\quad}$

(13) $\dfrac{1}{7} + \dfrac{1}{7} = \underline{\quad}$

(14) $\dfrac{1}{3} + \dfrac{2}{3} = \underline{\quad}$

(15) $\dfrac{2}{10} + \dfrac{4}{10} = \underline{\quad}$

(16) $\dfrac{5}{10} + \dfrac{5}{10} = \underline{\quad}$

(17) $\dfrac{3}{8} + \dfrac{5}{8} = \underline{\quad}$

(18) $\dfrac{4}{8} + \dfrac{1}{8} = \underline{\quad}$

(19) $\dfrac{2}{5} + \dfrac{2}{5} = \underline{\quad}$

(20) $\dfrac{2}{4} + \dfrac{3}{4} = \underline{\quad}$

(21) $\dfrac{1}{8} + \dfrac{3}{8} = \underline{\quad}$

(22) $\dfrac{3}{8} + \dfrac{3}{8} = \underline{\quad}$

(23) $\dfrac{1}{5} + \dfrac{1}{5} = \underline{\quad}$

(24) $\dfrac{1}{7} + \dfrac{6}{7} = \underline{\quad}$

(25) $\dfrac{4}{12} + \dfrac{4}{12} = \underline{\quad}$

(26) $\dfrac{4}{9} + \dfrac{4}{9} = \underline{\quad}$

(27) $\dfrac{2}{12} + \dfrac{3}{12} = \underline{\quad}$

(28) $\dfrac{1}{4} + \dfrac{1}{4} = \underline{\quad}$

(29) $\dfrac{2}{10} + \dfrac{4}{10} = \underline{\quad}$

(30) $\dfrac{3}{12} + \dfrac{1}{12} = \underline{\quad}$

(31) $\dfrac{5}{8} + \dfrac{5}{8} = \underline{\quad}$

(32) $\dfrac{1}{8} + \dfrac{3}{8} = \underline{\quad}$

(33) $\dfrac{1}{9} + \dfrac{3}{9} = \underline{\quad}$

(34) $\dfrac{2}{7} + \dfrac{2}{7} = \underline{\quad}$

(35) $\dfrac{3}{7} + \dfrac{3}{7} = \underline{\quad}$

(36) $\dfrac{2}{7} + \dfrac{5}{7} = \underline{\quad}$

(37) $\dfrac{3}{10} + \dfrac{3}{10} = \underline{\quad}$

(38) $\dfrac{2}{8} + \dfrac{4}{8} = \underline{\quad}$

(39) $\dfrac{1}{12} + \dfrac{3}{12} = \underline{\quad}$

(40) $\dfrac{4}{13} + \dfrac{4}{13} = \underline{\quad}$

(41) $\dfrac{1}{6} + \dfrac{3}{6} = \underline{\quad}$

(42) $\dfrac{1}{5} + \dfrac{3}{5} = \underline{\quad}$

(43) $\dfrac{1}{8} + \dfrac{1}{8} = \underline{\quad}$

(44) $\dfrac{4}{12} + \dfrac{5}{12} = \underline{\quad}$

(45) $\dfrac{2}{8} + \dfrac{5}{8} = \underline{\quad}$

Adding fractions with like denominators:

(1) $\dfrac{5}{12} + \dfrac{5}{12} = $ ——

(2) $\dfrac{2}{7} + \dfrac{2}{7} = $ ——

(3) $\dfrac{1}{6} + \dfrac{5}{6} = $ ——

(4) $\dfrac{2}{9} + \dfrac{2}{9} = $ ——

(5) $\dfrac{4}{10} + \dfrac{5}{10} = $ ——

(6) $\dfrac{2}{12} + \dfrac{5}{12} = $ ——

(7) $\dfrac{3}{12} + \dfrac{3}{12} = $ ——

(8) $\dfrac{1}{12} + \dfrac{1}{12} = $ ——

(9) $\dfrac{4}{10} + \dfrac{2}{10} = $ ——

(10) $\dfrac{4}{8} + \dfrac{4}{8} = $ ——

(11) $\dfrac{3}{9} + \dfrac{5}{9} = $ ——

(12) $\dfrac{3}{9} + \dfrac{6}{9} = $ ——

(13) $\dfrac{1}{6} + \dfrac{1}{6} = $ ——

(14) $\dfrac{2}{5} + \dfrac{3}{5} = $ ——

(15) $\dfrac{4}{7} + \dfrac{1}{7} = $ ——

(16) $\dfrac{5}{9} + \dfrac{5}{9} = $ ——

(17) $\dfrac{1}{8} + \dfrac{2}{8} = $ ——

(18) $\dfrac{3}{8} + \dfrac{5}{8} = $ ——

(19) $\dfrac{2}{6} + \dfrac{2}{6} = $ ——

(20) $\dfrac{4}{7} + \dfrac{4}{7} = $ ——

(21) $\dfrac{2}{6} + \dfrac{4}{6} = $ ——

(22) $\dfrac{3}{7} + \dfrac{3}{7} = $ ——

(23) $\dfrac{3}{8} + \dfrac{3}{8} = $ ——

(24) $\dfrac{1}{10} + \dfrac{4}{10} = $ ——

(25) $\dfrac{4}{10} + \dfrac{4}{10} = $ ——

(26) $\dfrac{2}{3} + \dfrac{4}{3} = $ ——

(27) $\dfrac{2}{5} + \dfrac{3}{5} = $ ——

(28) $\dfrac{1}{9} + \dfrac{1}{9} = $ ——

(29) $\dfrac{1}{4} + \dfrac{4}{4} = $ ——

(30) $\dfrac{4}{9} + \dfrac{1}{9} = $ ——

(31) $\dfrac{5}{12} + \dfrac{5}{12} = $ ——

(32) $\dfrac{3}{9} + \dfrac{6}{9} = $ ——

(33) $\dfrac{1}{12} + \dfrac{5}{12} = $ ——

(34) $\dfrac{2}{8} + \dfrac{2}{8} = $ ——

(35) $\dfrac{4}{8} + \dfrac{6}{8} = $ ——

(36) $\dfrac{3}{12} + \dfrac{2}{12} = $ ——

(37) $\dfrac{3}{12} + \dfrac{3}{12} = $ ——

(38) $\dfrac{1}{4} + \dfrac{4}{4} = $ ——

(39) $\dfrac{2}{9} + \dfrac{5}{9} = $ ——

(40) $\dfrac{4}{6} + \dfrac{4}{6} = $ ——

(41) $\dfrac{2}{7} + \dfrac{2}{7} = $ ——

(42) $\dfrac{1}{8} + \dfrac{2}{8} = $ ——

(43) $\dfrac{1}{7} + \dfrac{1}{7} = $ ——

(44) $\dfrac{1}{2} + \dfrac{3}{2} = $ ——

(45) $\dfrac{3}{6} + \dfrac{2}{6} = $ ——

Adding fractions with like denominators:

(1) $\dfrac{5}{10} + \dfrac{5}{10} =$ ⎯

(2) $\dfrac{2}{10} + \dfrac{5}{10} =$ ⎯

(3) $\dfrac{4}{12} + \dfrac{3}{12} =$ ⎯

(4) $\dfrac{2}{5} + \dfrac{2}{5} =$ ⎯

(5) $\dfrac{1}{8} + \dfrac{1}{8} =$ ⎯

(6) $\dfrac{2}{7} + \dfrac{4}{7} =$ ⎯

(7) $\dfrac{3}{8} + \dfrac{3}{8} =$ ⎯

(8) $\dfrac{3}{12} + \dfrac{3}{12} =$ ⎯

(9) $\dfrac{3}{10} + \dfrac{1}{10} =$ ⎯

(10) $\dfrac{4}{12} + \dfrac{4}{12} =$ ⎯

(11) $\dfrac{2}{3} + \dfrac{2}{3} =$ ⎯

(12) $\dfrac{4}{6} + \dfrac{1}{6} =$ ⎯

(13) $\dfrac{1}{12} + \dfrac{1}{12} =$ ⎯

(14) $\dfrac{3}{8} + \dfrac{2}{8} =$ ⎯

(15) $\dfrac{2}{8} + \dfrac{4}{8} =$ ⎯

(16) $\dfrac{5}{13} + \dfrac{5}{13} =$ ⎯

(17) $\dfrac{5}{6} + \dfrac{1}{6} =$ ⎯

(18) $\dfrac{1}{7} + \dfrac{4}{7} =$ ⎯

(19) $\dfrac{1}{2} + \dfrac{1}{2} =$ ⎯

(20) $\dfrac{1}{8} + \dfrac{5}{8} =$ ⎯

(21) $\dfrac{3}{12} + \dfrac{2}{12} =$ ⎯

(22) $\dfrac{2}{4} + \dfrac{2}{4} =$ ⎯

(23) $\dfrac{7}{9} + \dfrac{5}{9} =$ ⎯

(24) $\dfrac{1}{12} + \dfrac{4}{12} =$ ⎯

(25) $\dfrac{3}{6} + \dfrac{3}{6} =$ ⎯

(26) $\dfrac{3}{5} + \dfrac{2}{5} =$ ⎯

(27) $\dfrac{1}{6} + \dfrac{3}{6} =$ ⎯

(28) $\dfrac{4}{8} + \dfrac{4}{8} =$ ⎯

(29) $\dfrac{2}{6} + \dfrac{5}{6} =$ ⎯

(30) $\dfrac{2}{10} + \dfrac{3}{10} =$ ⎯

(31) $\dfrac{5}{10} + \dfrac{5}{10} =$ ⎯

(32) $\dfrac{1}{3} + \dfrac{2}{3} =$ ⎯

(33) $\dfrac{3}{9} + \dfrac{1}{9} =$ ⎯

(34) $\dfrac{2}{3} + \dfrac{2}{3} =$ ⎯

(35) $\dfrac{4}{7} + \dfrac{4}{7} =$ ⎯

(36) $\dfrac{1}{5} + \dfrac{4}{5} =$ ⎯

(37) $\dfrac{3}{9} + \dfrac{3}{9} =$ ⎯

(38) $\dfrac{3}{4} + \dfrac{1}{4} =$ ⎯

(39) $\dfrac{4}{12} + \dfrac{1}{12} =$ ⎯

(40) $\dfrac{4}{10} + \dfrac{4}{10} =$ ⎯

(41) $\dfrac{5}{8} + \dfrac{2}{8} =$ ⎯

(42) $\dfrac{2}{12} + \dfrac{3}{12} =$ ⎯

(43) $\dfrac{1}{8} + \dfrac{1}{8} =$ ⎯

(44) $\dfrac{2}{4} + \dfrac{4}{4} =$ ⎯

(45) $\dfrac{3}{7} + \dfrac{4}{7} =$ ⎯

Adding fractions with like denominators:

(1) $\dfrac{5}{15} + \dfrac{5}{15} = $ —

(2) $\dfrac{1}{9} + \dfrac{4}{9} = $ —

(3) $\dfrac{1}{9} + \dfrac{4}{9} = $ —

(4) $\dfrac{2}{5} + \dfrac{2}{5} = $ —

(5) $\dfrac{4}{10} + \dfrac{2}{10} = $ —

(6) $\dfrac{1}{8} + \dfrac{5}{8} = $ —

(7) $\dfrac{3}{7} + \dfrac{3}{7} = $ —

(8) $\dfrac{3}{8} + \dfrac{4}{8} = $ —

(9) $\dfrac{2}{6} + \dfrac{5}{6} = $ —

(10) $\dfrac{4}{9} + \dfrac{4}{9} = $ —

(11) $\dfrac{1}{6} + \dfrac{3}{6} = $ —

(12) $\dfrac{3}{12} + \dfrac{1}{12} = $ —

(13) $\dfrac{1}{3} + \dfrac{1}{3} = $ —

(14) $\dfrac{2}{8} + \dfrac{4}{8} = $ —

(15) $\dfrac{4}{8} + \dfrac{3}{8} = $ —

(16) $\dfrac{5}{8} + \dfrac{5}{8} = $ —

(17) $\dfrac{3}{6} + \dfrac{3}{6} = $ —

(18) $\dfrac{1}{10} + \dfrac{3}{10} = $ —

(19) $\dfrac{2}{4} + \dfrac{2}{4} = $ —

(20) $\dfrac{1}{7} + \dfrac{1}{7} = $ —

(21) $\dfrac{2}{7} + \dfrac{3}{7} = $ —

(22) $\dfrac{3}{5} + \dfrac{3}{5} = $ —

(23) $\dfrac{2}{5} + \dfrac{2}{5} = $ —

(24) $\dfrac{1}{12} + \dfrac{4}{12} = $ —

(25) $\dfrac{4}{7} + \dfrac{4}{7} = $ —

(26) $\dfrac{4}{9} + \dfrac{2}{9} = $ —

(27) $\dfrac{1}{5} + \dfrac{2}{5} = $ —

(28) $\dfrac{1}{10} + \dfrac{1}{10} = $ —

(29) $\dfrac{1}{5} + \dfrac{4}{5} = $ —

(30) $\dfrac{4}{9} + \dfrac{3}{9} = $ —

(31) $\dfrac{5}{13} + \dfrac{5}{13} = $ —

(32) $\dfrac{5}{7} + \dfrac{2}{7} = $ —

(33) $\dfrac{3}{8} + \dfrac{4}{8} = $ —

(34) $\dfrac{2}{6} + \dfrac{2}{6} = $ —

(35) $\dfrac{3}{10} + \dfrac{3}{10} = $ —

(36) $\dfrac{2}{12} + \dfrac{4}{12} = $ —

(37) $\dfrac{3}{8} + \dfrac{3}{8} = $ —

(38) $\dfrac{2}{12} + \dfrac{4}{12} = $ —

(39) $\dfrac{4}{7} + \dfrac{3}{7} = $ —

(40) $\dfrac{4}{12} + \dfrac{4}{12} = $ —

(41) $\dfrac{1}{10} + \dfrac{3}{10} = $ —

(42) $\dfrac{1}{9} + \dfrac{2}{9} = $ —

(43) $\dfrac{1}{12} + \dfrac{1}{12} = $ —

(44) $\dfrac{4}{6} + \dfrac{2}{6} = $ —

(45) $\dfrac{2}{5} + \dfrac{4}{5} = $ —

Subtracting fractions with like denominators:

(1) $\dfrac{6}{7} - \dfrac{5}{7} =$ ___

(2) $\dfrac{3}{7} - \dfrac{1}{7} =$ ___

(3) $\dfrac{3}{4} - \dfrac{2}{4} =$ ___

(4) $\dfrac{7}{8} - \dfrac{2}{8} =$ ___

(5) $\dfrac{4}{9} - \dfrac{2}{9} =$ ___

(6) $\dfrac{2}{6} - \dfrac{1}{6} =$ ___

(7) $\dfrac{8}{10} - \dfrac{3}{10} =$ ___

(8) $\dfrac{3}{12} - \dfrac{1}{12} =$ ___

(9) $\dfrac{5}{9} - \dfrac{4}{9} =$ ___

(10) $\dfrac{5}{12} - \dfrac{4}{12} =$ ___

(11) $\dfrac{5}{8} - \dfrac{3}{8} =$ ___

(12) $\dfrac{7}{10} - \dfrac{4}{10} =$ ___

(13) $\dfrac{8}{5} - \dfrac{1}{5} =$ ___

(14) $\dfrac{6}{8} - \dfrac{2}{8} =$ ___

(15) $\dfrac{2}{3} - \dfrac{1}{3} =$ ___

(16) $\dfrac{6}{9} - \dfrac{5}{9} =$ ___

(17) $\dfrac{2}{8} - \dfrac{1}{8} =$ ___

(18) $\dfrac{2}{2} - \dfrac{1}{2} =$ ___

(19) $\dfrac{8}{9} - \dfrac{2}{9} =$ ___

(20) $\dfrac{7}{8} - \dfrac{4}{8} =$ ___

(21) $\dfrac{7}{10} - \dfrac{3}{10} =$ ___

(22) $\dfrac{5}{12} - \dfrac{3}{12} =$ ___

(23) $\dfrac{4}{9} - \dfrac{3}{9} =$ ___

(24) $\dfrac{5}{8} - \dfrac{2}{8} =$ ___

(25) $\dfrac{6}{6} - \dfrac{4}{6} =$ ___

(26) $\dfrac{3}{3} - \dfrac{1}{3} =$ ___

(27) $\dfrac{4}{7} - \dfrac{2}{7} =$ ___

(28) $\dfrac{6}{12} - \dfrac{1}{12} =$ ___

(29) $\dfrac{3}{4} - \dfrac{2}{4} =$ ___

(30) $\dfrac{3}{4} - \dfrac{1}{4} =$ ___

(31) $\dfrac{7}{10} - \dfrac{5}{10} =$ ___

(32) $\dfrac{4}{5} - \dfrac{2}{5} =$ ___

(33) $\dfrac{3}{5} - \dfrac{2}{5} =$ ___

(34) $\dfrac{6}{12} - \dfrac{2}{12} =$ ___

(35) $\dfrac{5}{7} - \dfrac{3}{7} =$ ___

(36) $\dfrac{4}{7} - \dfrac{2}{7} =$ ___

(37) $\dfrac{7}{7} - \dfrac{3}{7} =$ ___

(38) $\dfrac{5}{6} - \dfrac{1}{6} =$ ___

(39) $\dfrac{5}{8} - \dfrac{3}{8} =$ ___

(40) $\dfrac{5}{8} - \dfrac{4}{8} =$ ___

(41) $\dfrac{5}{7} - \dfrac{2}{7} =$ ___

(42) $\dfrac{4}{3} - \dfrac{1}{3} =$ ___

(43) $\dfrac{7}{7} - \dfrac{1}{7} =$ ___

(44) $\dfrac{9}{12} - \dfrac{4}{12} =$ ___

(45) $\dfrac{5}{6} - \dfrac{4}{6} =$ ___

Subtracting fractions with like denominators:

(1) $\dfrac{6}{6} - \dfrac{5}{6} =$ —

(2) $\dfrac{4}{4} - \dfrac{3}{4} =$ —

(3) $\dfrac{3}{4} - \dfrac{2}{4} =$ —

(4) $\dfrac{6}{10} - \dfrac{2}{10} =$ —

(5) $\dfrac{7}{8} - \dfrac{1}{8} =$ —

(6) $\dfrac{7}{8} - \dfrac{2}{8} =$ —

(7) $\dfrac{8}{12} - \dfrac{3}{12} =$ —

(8) $\dfrac{7}{8} - \dfrac{2}{8} =$ —

(9) $\dfrac{5}{9} - \dfrac{1}{9} =$ —

(10) $\dfrac{9}{9} - \dfrac{4}{9} =$ —

(11) $\dfrac{5}{9} - \dfrac{1}{9} =$ —

(12) $\dfrac{5}{10} - \dfrac{3}{10} =$ —

(13) $\dfrac{6}{4} - \dfrac{1}{4} =$ —

(14) $\dfrac{2}{5} - \dfrac{1}{5} =$ —

(15) $\dfrac{3}{5} - \dfrac{1}{5} =$ —

(16) $\dfrac{8}{8} - \dfrac{5}{8} =$ —

(17) $\dfrac{5}{10} - \dfrac{3}{10} =$ —

(18) $\dfrac{4}{6} - \dfrac{2}{6} =$ —

(19) $\dfrac{8}{5} - \dfrac{2}{5} =$ —

(20) $\dfrac{3}{3} - \dfrac{2}{3} =$ —

(21) $\dfrac{5}{8} - \dfrac{2}{8} =$ —

(22) $\dfrac{6}{6} - \dfrac{3}{6} =$ —

(23) $\dfrac{4}{7} - \dfrac{1}{7} =$ —

(24) $\dfrac{4}{2} - \dfrac{1}{2} =$ —

(25) $\dfrac{5}{7} - \dfrac{4}{7} =$ —

(26) $\dfrac{2}{9} - \dfrac{1}{9} =$ —

(27) $\dfrac{7}{12} - \dfrac{4}{12} =$ —

(28) $\dfrac{6}{10} - \dfrac{1}{10} =$ —

(29) $\dfrac{3}{6} - \dfrac{2}{6} =$ —

(30) $\dfrac{3}{7} - \dfrac{2}{7} =$ —

(31) $\dfrac{8}{12} - \dfrac{5}{12} =$ —

(32) $\dfrac{8}{12} - \dfrac{2}{12} =$ —

(33) $\dfrac{3}{6} - \dfrac{1}{6} =$ —

(34) $\dfrac{9}{7} - \dfrac{2}{7} =$ —

(35) $\dfrac{10}{10} - \dfrac{4}{10} =$ —

(36) $\dfrac{5}{10} - \dfrac{3}{10} =$ —

(37) $\dfrac{6}{9} - \dfrac{3}{9} =$ —

(38) $\dfrac{2}{6} - \dfrac{1}{6} =$ —

(39) $\dfrac{2}{3} - \dfrac{1}{3} =$ —

(40) $\dfrac{7}{12} - \dfrac{4}{12} =$ —

(41) $\dfrac{3}{4} - \dfrac{1}{4} =$ —

(42) $\dfrac{7}{9} - \dfrac{4}{9} =$ —

(43) $\dfrac{8}{4} - \dfrac{1}{4} =$ —

(44) $\dfrac{4}{9} - \dfrac{2}{9} =$ —

(45) $\dfrac{4}{8} - \dfrac{3}{8} =$ —

Subtracting fractions with like denominators:

(1) $\dfrac{8}{9} - \dfrac{5}{9} = \underline{\quad}$

(2) $\dfrac{3}{3} - \dfrac{2}{3} = \underline{\quad}$

(3) $\dfrac{4}{10} - \dfrac{2}{10} = \underline{\quad}$

(4) $\dfrac{7}{6} - \dfrac{2}{6} = \underline{\quad}$

(5) $\dfrac{6}{7} - \dfrac{1}{7} = \underline{\quad}$

(6) $\dfrac{3}{4} - \dfrac{1}{4} = \underline{\quad}$

(7) $\dfrac{8}{8} - \dfrac{3}{8} = \underline{\quad}$

(8) $\dfrac{6}{8} - \dfrac{5}{8} = \underline{\quad}$

(9) $\dfrac{3}{8} - \dfrac{1}{8} = \underline{\quad}$

(10) $\dfrac{6}{10} - \dfrac{4}{10} = \underline{\quad}$

(11) $\dfrac{5}{8} - \dfrac{3}{8} = \underline{\quad}$

(12) $\dfrac{6}{10} - \dfrac{2}{10} = \underline{\quad}$

(13) $\dfrac{7}{5} - \dfrac{1}{5} = \underline{\quad}$

(14) $\dfrac{8}{8} - \dfrac{1}{8} = \underline{\quad}$

(15) $\dfrac{2}{7} - \dfrac{1}{7} = \underline{\quad}$

(16) $\dfrac{4}{6} - \dfrac{1}{6} = \underline{\quad}$

(17) $\dfrac{5}{5} - \dfrac{1}{5} = \underline{\quad}$

(18) $\dfrac{4}{6} - \dfrac{2}{6} = \underline{\quad}$

(19) $\dfrac{3}{4} - \dfrac{1}{4} = \underline{\quad}$

(20) $\dfrac{5}{5} - \dfrac{3}{5} = \underline{\quad}$

(21) $\dfrac{5}{12} - \dfrac{4}{12} = \underline{\quad}$

(22) $\dfrac{2}{8} - \dfrac{1}{8} = \underline{\quad}$

(23) $\dfrac{6}{12} - \dfrac{1}{12} = \underline{\quad}$

(24) $\dfrac{2}{9} - \dfrac{1}{9} = \underline{\quad}$

(25) $\dfrac{2}{10} - \dfrac{1}{10} = \underline{\quad}$

(26) $\dfrac{6}{6} - \dfrac{5}{6} = \underline{\quad}$

(27) $\dfrac{3}{5} - \dfrac{2}{5} = \underline{\quad}$

(28) $\dfrac{4}{9} - \dfrac{2}{9} = \underline{\quad}$

(29) $\dfrac{6}{10} - \dfrac{3}{10} = \underline{\quad}$

(30) $\dfrac{6}{7} - \dfrac{1}{7} = \underline{\quad}$

(31) $\dfrac{5}{7} - \dfrac{1}{7} = \underline{\quad}$

(32) $\dfrac{5}{5} - \dfrac{4}{5} = \underline{\quad}$

(33) $\dfrac{2}{8} - \dfrac{1}{8} = \underline{\quad}$

(34) $\dfrac{5}{8} - \dfrac{3}{8} = \underline{\quad}$

(35) $\dfrac{4}{8} - \dfrac{1}{8} = \underline{\quad}$

(36) $\dfrac{6}{10} - \dfrac{4}{10} = \underline{\quad}$

(37) $\dfrac{3}{12} - \dfrac{2}{12} = \underline{\quad}$

(38) $\dfrac{9}{9} - \dfrac{4}{9} = \underline{\quad}$

(39) $\dfrac{3}{4} - \dfrac{1}{4} = \underline{\quad}$

(40) $\dfrac{5}{12} - \dfrac{1}{12} = \underline{\quad}$

(41) $\dfrac{4}{4} - \dfrac{2}{4} = \underline{\quad}$

(42) $\dfrac{5}{9} - \dfrac{3}{9} = \underline{\quad}$

(43) $\dfrac{5}{12} - \dfrac{3}{12} = \underline{\quad}$

(44) $\dfrac{4}{10} - \dfrac{1}{10} = \underline{\quad}$

(45) $\dfrac{2}{10} - \dfrac{1}{10} = \underline{\quad}$

Subtracting fractions with like denominators:

(1) $\dfrac{3}{5} - \dfrac{2}{5} = $ ___

(2) $\dfrac{4}{12} - \dfrac{2}{12} = $ ___

(3) $\dfrac{3}{6} - \dfrac{2}{6} = $ ___

(4) $\dfrac{5}{9} - \dfrac{1}{9} = $ ___

(5) $\dfrac{4}{3} - \dfrac{1}{3} = $ ___

(6) $\dfrac{4}{5} - \dfrac{1}{5} = $ ___

(7) $\dfrac{5}{7} - \dfrac{2}{7} = $ ___

(8) $\dfrac{6}{7} - \dfrac{3}{7} = $ ___

(9) $\dfrac{7}{12} - \dfrac{4}{12} = $ ___

(10) $\dfrac{4}{12} - \dfrac{1}{12} = $ ___

(11) $\dfrac{8}{8} - \dfrac{2}{8} = $ ___

(12) $\dfrac{4}{7} - \dfrac{1}{7} = $ ___

(13) $\dfrac{3}{4} - \dfrac{2}{4} = $ ___

(14) $\dfrac{3}{9} - \dfrac{1}{9} = $ ___

(15) $\dfrac{4}{9} - \dfrac{2}{9} = $ ___

(16) $\dfrac{4}{10} - \dfrac{1}{10} = $ ___

(17) $\dfrac{3}{9} - \dfrac{1}{9} = $ ___

(18) $\dfrac{3}{10} - \dfrac{2}{10} = $ ___

(19) $\dfrac{4}{6} - \dfrac{2}{6} = $ ___

(20) $\dfrac{5}{8} - \dfrac{3}{8} = $ ___

(21) $\dfrac{4}{10} - \dfrac{3}{10} = $ ___

(22) $\dfrac{4}{10} - \dfrac{3}{10} = $ ___

(23) $\dfrac{7}{10} - \dfrac{2}{10} = $ ___

(24) $\dfrac{5}{7} - \dfrac{3}{7} = $ ___

(25) $\dfrac{4}{5} - \dfrac{1}{5} = $ ___

(26) $\dfrac{10}{12} - \dfrac{1}{12} = $ ___

(27) $\dfrac{2}{3} - \dfrac{1}{3} = $ ___

(28) $\dfrac{4}{8} - \dfrac{2}{8} = $ ___

(29) $\dfrac{8}{8} - \dfrac{1}{8} = $ ___

(30) $\dfrac{2}{4} - \dfrac{1}{4} = $ ___

(31) $\dfrac{4}{7} - \dfrac{1}{7} = $ ___

(32) $\dfrac{9}{6} - \dfrac{3}{6} = $ ___

(33) $\dfrac{3}{8} - \dfrac{2}{8} = $ ___

(34) $\dfrac{4}{12} - \dfrac{2}{12} = $ ___

(35) $\dfrac{3}{3} - \dfrac{1}{3} = $ ___

(36) $\dfrac{4}{7} - \dfrac{1}{7} = $ ___

(37) $\dfrac{5}{9} - \dfrac{3}{9} = $ ___

(38) $\dfrac{8}{10} - \dfrac{2}{10} = $ ___

(39) $\dfrac{12}{10} - \dfrac{5}{10} = $ ___

(40) $\dfrac{5}{8} - \dfrac{1}{8} = $ ___

(41) $\dfrac{6}{7} - \dfrac{4}{7} = $ ___

(42) $\dfrac{2}{5} - \dfrac{1}{5} = $ ___

(43) $\dfrac{5}{3} - \dfrac{2}{3} = $ ___

(44) $\dfrac{6}{12} - \dfrac{1}{12} = $ ___

(45) $\dfrac{6}{9} - \dfrac{3}{9} = $ ___

Subtracting fractions with like denominators:

(1) $\dfrac{5}{5} - \dfrac{3}{5} = $ —

(2) $\dfrac{6}{8} - \dfrac{4}{8} = $ —

(3) $\dfrac{3}{6} - \dfrac{2}{6} = $ —

(4) $\dfrac{3}{10} - \dfrac{2}{10} = $ —

(5) $\dfrac{6}{4} - \dfrac{1}{4} = $ —

(6) $\dfrac{3}{9} - \dfrac{1}{9} = $ —

(7) $\dfrac{5}{3} - \dfrac{1}{3} = $ —

(8) $\dfrac{3}{7} - \dfrac{2}{7} = $ —

(9) $\dfrac{5}{12} - \dfrac{2}{12} = $ —

(10) $\dfrac{5}{9} - \dfrac{2}{9} = $ —

(11) $\dfrac{4}{6} - \dfrac{2}{6} = $ —

(12) $\dfrac{3}{2} - \dfrac{1}{2} = $ —

(13) $\dfrac{3}{6} - \dfrac{1}{6} = $ —

(14) $\dfrac{8}{10} - \dfrac{1}{10} = $ —

(15) $\dfrac{7}{8} - \dfrac{3}{8} = $ —

(16) $\dfrac{4}{12} - \dfrac{2}{12} = $ —

(17) $\dfrac{7}{10} - \dfrac{4}{10} = $ —

(18) $\dfrac{4}{6} - \dfrac{1}{6} = $ —

(19) $\dfrac{3}{12} - \dfrac{1}{12} = $ —

(20) $\dfrac{7}{12} - \dfrac{3}{12} = $ —

(21) $\dfrac{5}{7} - \dfrac{4}{7} = $ —

(22) $\dfrac{3}{8} - \dfrac{2}{8} = $ —

(23) $\dfrac{3}{6} - \dfrac{1}{6} = $ —

(24) $\dfrac{4}{5} - \dfrac{2}{5} = $ —

(25) $\dfrac{5}{4} - \dfrac{1}{4} = $ —

(26) $\dfrac{7}{4} - \dfrac{3}{4} = $ —

(27) $\dfrac{3}{4} - \dfrac{1}{4} = $ —

(28) $\dfrac{5}{7} - \dfrac{3}{7} = $ —

(29) $\dfrac{3}{5} - \dfrac{2}{5} = $ —

(30) $\dfrac{6}{9} - \dfrac{3}{9} = $ —

(31) $\dfrac{2}{9} - \dfrac{1}{9} = $ —

(32) $\dfrac{9}{9} - \dfrac{3}{9} = $ —

(33) $\dfrac{4}{7} - \dfrac{2}{7} = $ —

(34) $\dfrac{5}{6} - \dfrac{2}{6} = $ —

(35) $\dfrac{11}{12} - \dfrac{1}{12} = $ —

(36) $\dfrac{2}{3} - \dfrac{1}{3} = $ —

(37) $\dfrac{5}{12} - \dfrac{3}{12} = $ —

(38) $\dfrac{3}{8} - \dfrac{2}{8} = $ —

(39) $\dfrac{5}{10} - \dfrac{4}{10} = $ —

(40) $\dfrac{5}{5} - \dfrac{1}{5} = $ —

(41) $\dfrac{9}{9} - \dfrac{1}{9} = $ —

(42) $\dfrac{3}{5} - \dfrac{1}{5} = $ —

(43) $\dfrac{3}{7} - \dfrac{2}{7} = $ —

(44) $\dfrac{7}{5} - \dfrac{1}{5} = $ —

(45) $\dfrac{7}{12} - \dfrac{5}{12} = $ —

Subtracting fractions with like denominators:

(1) $\dfrac{5}{12} - \dfrac{1}{12} = $ ―

(2) $\dfrac{9}{9} - \dfrac{2}{9} = $ ―

(3) $\dfrac{3}{6} - \dfrac{2}{6} = $ ―

(4) $\dfrac{5}{6} - \dfrac{3}{6} = $ ―

(5) $\dfrac{3}{8} - \dfrac{1}{8} = $ ―

(6) $\dfrac{5}{10} - \dfrac{4}{10} = $ ―

(7) $\dfrac{5}{4} - \dfrac{2}{4} = $ ―

(8) $\dfrac{3}{12} - \dfrac{2}{12} = $ ―

(9) $\dfrac{2}{5} - \dfrac{1}{5} = $ ―

(10) $\dfrac{5}{10} - \dfrac{3}{10} = $ ―

(11) $\dfrac{7}{5} - \dfrac{3}{5} = $ ―

(12) $\dfrac{4}{8} - \dfrac{2}{8} = $ ―

(13) $\dfrac{4}{8} - \dfrac{1}{8} = $ ―

(14) $\dfrac{5}{4} - \dfrac{1}{4} = $ ―

(15) $\dfrac{4}{7} - \dfrac{3}{7} = $ ―

(16) $\dfrac{4}{5} - \dfrac{2}{5} = $ ―

(17) $\dfrac{7}{8} - \dfrac{1}{8} = $ ―

(18) $\dfrac{6}{8} - \dfrac{2}{8} = $ ―

(19) $\dfrac{5}{10} - \dfrac{1}{10} = $ ―

(20) $\dfrac{3}{4} - \dfrac{2}{4} = $ ―

(21) $\dfrac{5}{9} - \dfrac{2}{9} = $ ―

(22) $\dfrac{3}{12} - \dfrac{2}{12} = $ ―

(23) $\dfrac{7}{5} - \dfrac{4}{5} = $ ―

(24) $\dfrac{5}{7} - \dfrac{1}{7} = $ ―

(25) $\dfrac{3}{7} - \dfrac{1}{7} = $ ―

(26) $\dfrac{10}{10} - \dfrac{1}{10} = $ ―

(27) $\dfrac{3}{4} - \dfrac{1}{4} = $ ―

(28) $\dfrac{4}{3} - \dfrac{2}{3} = $ ―

(29) $\dfrac{5}{6} - \dfrac{4}{6} = $ ―

(30) $\dfrac{5}{7} - \dfrac{4}{7} = $ ―

(31) $\dfrac{4}{5} - \dfrac{3}{5} = $ ―

(32) $\dfrac{5}{7} - \dfrac{2}{7} = $ ―

(33) $\dfrac{4}{4} - \dfrac{2}{4} = $ ―

(34) $\dfrac{4}{12} - \dfrac{1}{12} = $ ―

(35) $\dfrac{5}{7} - \dfrac{2}{7} = $ ―

(36) $\dfrac{2}{6} - \dfrac{1}{6} = $ ―

(37) $\dfrac{4}{8} - \dfrac{3}{8} = $ ―

(38) $\dfrac{5}{9} - \dfrac{2}{9} = $ ―

(39) $\dfrac{4}{10} - \dfrac{3}{10} = $ ―

(40) $\dfrac{5}{12} - \dfrac{2}{12} = $ ―

(41) $\dfrac{7}{8} - \dfrac{1}{8} = $ ―

(42) $\dfrac{5}{6} - \dfrac{3}{6} = $ ―

(43) $\dfrac{4}{4} - \dfrac{1}{4} = $ ―

(44) $\dfrac{4}{6} - \dfrac{3}{6} = $ ―

(45) $\dfrac{4}{8} - \dfrac{1}{8} = $ ―

Subtracting fractions with like denominators:

(1) $\dfrac{4}{10} - \dfrac{2}{10} = \underline{\quad}$

(2) $\dfrac{7}{10} - \dfrac{4}{10} = \underline{\quad}$

(3) $\dfrac{7}{9} - \dfrac{2}{9} = \underline{\quad}$

(4) $\dfrac{5}{6} - \dfrac{1}{6} = \underline{\quad}$

(5) $\dfrac{4}{10} - \dfrac{2}{10} = \underline{\quad}$

(6) $\dfrac{8}{10} - \dfrac{2}{10} = \underline{\quad}$

(7) $\dfrac{4}{9} - \dfrac{3}{9} = \underline{\quad}$

(8) $\dfrac{8}{10} - \dfrac{3}{10} = \underline{\quad}$

(9) $\dfrac{6}{10} - \dfrac{4}{10} = \underline{\quad}$

(10) $\dfrac{5}{8} - \dfrac{2}{8} = \underline{\quad}$

(11) $\dfrac{4}{12} - \dfrac{1}{12} = \underline{\quad}$

(12) $\dfrac{5}{8} - \dfrac{3}{8} = \underline{\quad}$

(13) $\dfrac{3}{5} - \dfrac{1}{5} = \underline{\quad}$

(14) $\dfrac{3}{10} - \dfrac{2}{10} = \underline{\quad}$

(15) $\dfrac{2}{3} - \dfrac{1}{3} = \underline{\quad}$

(16) $\dfrac{8}{8} - \dfrac{2}{8} = \underline{\quad}$

(17) $\dfrac{9}{12} - \dfrac{2}{12} = \underline{\quad}$

(18) $\dfrac{3}{3} - \dfrac{2}{3} = \underline{\quad}$

(19) $\dfrac{7}{10} - \dfrac{3}{10} = \underline{\quad}$

(20) $\dfrac{4}{7} - \dfrac{1}{7} = \underline{\quad}$

(21) $\dfrac{5}{12} - \dfrac{3}{12} = \underline{\quad}$

(22) $\dfrac{3}{5} - \dfrac{2}{5} = \underline{\quad}$

(23) $\dfrac{5}{8} - \dfrac{3}{8} = \underline{\quad}$

(24) $\dfrac{4}{9} - \dfrac{1}{9} = \underline{\quad}$

(25) $\dfrac{7}{9} - \dfrac{1}{9} = \underline{\quad}$

(26) $\dfrac{7}{10} - \dfrac{1}{10} = \underline{\quad}$

(27) $\dfrac{2}{10} - \dfrac{1}{10} = \underline{\quad}$

(28) $\dfrac{4}{12} - \dfrac{3}{12} = \underline{\quad}$

(29) $\dfrac{3}{5} - \dfrac{2}{5} = \underline{\quad}$

(30) $\dfrac{4}{7} - \dfrac{3}{7} = \underline{\quad}$

(31) $\dfrac{7}{8} - \dfrac{2}{8} = \underline{\quad}$

(32) $\dfrac{8}{9} - \dfrac{1}{9} = \underline{\quad}$

(33) $\dfrac{3}{9} - \dfrac{2}{9} = \underline{\quad}$

(34) $\dfrac{5}{6} - \dfrac{1}{6} = \underline{\quad}$

(35) $\dfrac{3}{3} - \dfrac{2}{3} = \underline{\quad}$

(36) $\dfrac{5}{10} - \dfrac{4}{10} = \underline{\quad}$

(37) $\dfrac{7}{10} - \dfrac{3}{10} = \underline{\quad}$

(38) $\dfrac{3}{5} - \dfrac{1}{5} = \underline{\quad}$

(39) $\dfrac{2}{4} - \dfrac{1}{4} = \underline{\quad}$

(40) $\dfrac{5}{7} - \dfrac{2}{7} = \underline{\quad}$

(41) $\dfrac{9}{10} - \dfrac{1}{10} = \underline{\quad}$

(42) $\dfrac{4}{5} - \dfrac{3}{5} = \underline{\quad}$

(43) $\dfrac{5}{8} - \dfrac{1}{8} = \underline{\quad}$

(44) $\dfrac{6}{8} - \dfrac{2}{8} = \underline{\quad}$

(45) $\dfrac{5}{8} - \dfrac{4}{8} = \underline{\quad}$

Subtracting fractions with like denominators:

(1) $\dfrac{5}{12} - \dfrac{2}{12} = $ —

(2) $\dfrac{2}{3} - \dfrac{1}{3} = $ —

(3) $\dfrac{3}{10} - \dfrac{2}{10} = $ —

(4) $\dfrac{4}{5} - \dfrac{1}{5} = $ —

(5) $\dfrac{3}{7} - \dfrac{2}{7} = $ —

(6) $\dfrac{3}{7} - \dfrac{1}{7} = $ —

(7) $\dfrac{5}{4} - \dfrac{3}{4} = $ —

(8) $\dfrac{5}{12} - \dfrac{3}{12} = $ —

(9) $\dfrac{6}{10} - \dfrac{4}{10} = $ —

(10) $\dfrac{5}{12} - \dfrac{2}{12} = $ —

(11) $\dfrac{6}{8} - \dfrac{4}{8} = $ —

(12) $\dfrac{3}{5} - \dfrac{2}{5} = $ —

(13) $\dfrac{4}{7} - \dfrac{1}{7} = $ —

(14) $\dfrac{3}{4} - \dfrac{1}{4} = $ —

(15) $\dfrac{6}{12} - \dfrac{4}{12} = $ —

(16) $\dfrac{4}{6} - \dfrac{3}{6} = $ —

(17) $\dfrac{4}{4} - \dfrac{2}{4} = $ —

(18) $\dfrac{6}{9} - \dfrac{3}{9} = $ —

(19) $\dfrac{3}{10} - \dfrac{2}{10} = $ —

(20) $\dfrac{4}{6} - \dfrac{2}{6} = $ —

(21) $\dfrac{6}{10} - \dfrac{4}{10} = $ —

(22) $\dfrac{4}{12} - \dfrac{1}{12} = $ —

(23) $\dfrac{3}{2} - \dfrac{1}{2} = $ —

(24) $\dfrac{4}{4} - \dfrac{1}{4} = $ —

(25) $\dfrac{6}{4} - \dfrac{2}{4} = $ —

(26) $\dfrac{4}{5} - \dfrac{3}{5} = $ —

(27) $\dfrac{5}{7} - \dfrac{2}{7} = $ —

(28) $\dfrac{6}{9} - \dfrac{1}{9} = $ —

(29) $\dfrac{5}{6} - \dfrac{1}{6} = $ —

(30) $\dfrac{3}{2} - \dfrac{1}{2} = $ —

(31) $\dfrac{4}{7} - \dfrac{3}{7} = $ —

(32) $\dfrac{6}{12} - \dfrac{1}{12} = $ —

(33) $\dfrac{5}{10} - \dfrac{2}{10} = $ —

(34) $\dfrac{4}{9} - \dfrac{2}{9} = $ —

(35) $\dfrac{4}{7} - \dfrac{3}{7} = $ —

(36) $\dfrac{7}{8} - \dfrac{1}{8} = $ —

(37) $\dfrac{3}{4} - \dfrac{1}{4} = $ —

(38) $\dfrac{8}{9} - \dfrac{2}{9} = $ —

(39) $\dfrac{9}{12} - \dfrac{3}{12} = $ —

(40) $\dfrac{5}{5} - \dfrac{3}{5} = $ —

(41) $\dfrac{9}{8} - \dfrac{1}{8} = $ —

(42) $\dfrac{3}{3} - \dfrac{2}{3} = $ —

(43) $\dfrac{3}{6} - \dfrac{2}{6} = $ —

(44) $\dfrac{3}{10} - \dfrac{2}{10} = $ —

(45) $\dfrac{3}{8} - \dfrac{2}{8} = $ —

Day:	9		Date:		Score:	/45
Name:			Time:	:	Rating:	☆☆☆☆☆

Subtracting fractions with like denominators:

(1) $\dfrac{5}{10} - \dfrac{1}{10} = \underline{}$

(2) $\dfrac{10}{10} - \dfrac{1}{10} = \underline{}$

(3) $\dfrac{5}{6} - \dfrac{1}{6} = \underline{}$

(4) $\dfrac{5}{8} - \dfrac{2}{8} = \underline{}$

(5) $\dfrac{3}{7} - \dfrac{1}{7} = \underline{}$

(6) $\dfrac{5}{10} - \dfrac{1}{10} = \underline{}$

(7) $\dfrac{5}{12} - \dfrac{1}{12} = \underline{}$

(8) $\dfrac{7}{10} - \dfrac{2}{10} = \underline{}$

(9) $\dfrac{7}{9} - \dfrac{5}{9} = \underline{}$

(10) $\dfrac{5}{8} - \dfrac{3}{8} = \underline{}$

(11) $\dfrac{9}{10} - \dfrac{1}{10} = \underline{}$

(12) $\dfrac{3}{5} - \dfrac{2}{5} = \underline{}$

(13) $\dfrac{3}{12} - \dfrac{2}{12} = \underline{}$

(14) $\dfrac{4}{5} - \dfrac{2}{5} = \underline{}$

(15) $\dfrac{5}{6} - \dfrac{2}{6} = \underline{}$

(16) $\dfrac{3}{7} - \dfrac{1}{7} = \underline{}$

(17) $\dfrac{9}{9} - \dfrac{1}{9} = \underline{}$

(18) $\dfrac{2}{3} - \dfrac{1}{3} = \underline{}$

(19) $\dfrac{6}{9} - \dfrac{3}{9} = \underline{}$

(20) $\dfrac{4}{3} - \dfrac{2}{3} = \underline{}$

(21) $\dfrac{5}{7} - \dfrac{4}{7} = \underline{}$

(22) $\dfrac{6}{10} - \dfrac{2}{10} = \underline{}$

(23) $\dfrac{4}{8} - \dfrac{3}{8} = \underline{}$

(24) $\dfrac{3}{4} - \dfrac{1}{4} = \underline{}$

(25) $\dfrac{3}{6} - \dfrac{1}{6} = \underline{}$

(26) $\dfrac{4}{5} - \dfrac{1}{5} = \underline{}$

(27) $\dfrac{5}{8} - \dfrac{2}{8} = \underline{}$

(28) $\dfrac{5}{12} - \dfrac{3}{12} = \underline{}$

(29) $\dfrac{4}{4} - \dfrac{1}{4} = \underline{}$

(30) $\dfrac{4}{4} - \dfrac{2}{4} = \underline{}$

(31) $\dfrac{3}{7} - \dfrac{2}{7} = \underline{}$

(32) $\dfrac{7}{12} - \dfrac{1}{12} = \underline{}$

(33) $\dfrac{4}{9} - \dfrac{1}{9} = \underline{}$

(34) $\dfrac{3}{8} - \dfrac{1}{8} = \underline{}$

(35) $\dfrac{4}{7} - \dfrac{2}{7} = \underline{}$

(36) $\dfrac{4}{10} - \dfrac{2}{10} = \underline{}$

(37) $\dfrac{5}{6} - \dfrac{2}{6} = \underline{}$

(38) $\dfrac{5}{12} - \dfrac{2}{12} = \underline{}$

(39) $\dfrac{4}{10} - \dfrac{3}{10} = \underline{}$

(40) $\dfrac{3}{12} - \dfrac{1}{12} = \underline{}$

(41) $\dfrac{4}{3} - \dfrac{1}{3} = \underline{}$

(42) $\dfrac{3}{6} - \dfrac{1}{6} = \underline{}$

(43) $\dfrac{6}{4} - \dfrac{3}{4} = \underline{}$

(44) $\dfrac{3}{4} - \dfrac{2}{4} = \underline{}$

(45) $\dfrac{4}{8} - \dfrac{2}{8} = \underline{}$

Subtracting fractions with like denominators:

(1) $\dfrac{4}{12} - \dfrac{2}{12} = $ —

(2) $\dfrac{6}{6} - \dfrac{1}{6} = $ —

(3) $\dfrac{6}{6} - \dfrac{3}{6} = $ —

(4) $\dfrac{5}{7} - \dfrac{1}{7} = $ —

(5) $\dfrac{5}{5} - \dfrac{3}{5} = $ —

(6) $\dfrac{4}{7} - \dfrac{1}{7} = $ —

(7) $\dfrac{5}{6} - \dfrac{3}{6} = $ —

(8) $\dfrac{8}{10} - \dfrac{1}{10} = $ —

(9) $\dfrac{5}{5} - \dfrac{2}{5} = $ —

(10) $\dfrac{3}{9} - \dfrac{2}{9} = $ —

(11) $\dfrac{7}{8} - \dfrac{2}{8} = $ —

(12) $\dfrac{4}{9} - \dfrac{2}{9} = $ —

(13) $\dfrac{3}{10} - \dfrac{1}{10} = $ —

(14) $\dfrac{5}{7} - \dfrac{1}{7} = $ —

(15) $\dfrac{4}{5} - \dfrac{1}{5} = $ —

(16) $\dfrac{6}{5} - \dfrac{3}{5} = $ —

(17) $\dfrac{4}{12} - \dfrac{3}{12} = $ —

(18) $\dfrac{5}{7} - \dfrac{2}{7} = $ —

(19) $\dfrac{4}{8} - \dfrac{2}{8} = $ —

(20) $\dfrac{6}{9} - \dfrac{2}{9} = $ —

(21) $\dfrac{7}{10} - \dfrac{3}{10} = $ —

(22) $\dfrac{3}{5} - \dfrac{1}{5} = $ —

(23) $\dfrac{10}{8} - \dfrac{1}{8} = $ —

(24) $\dfrac{4}{12} - \dfrac{2}{12} = $ —

(25) $\dfrac{5}{7} - \dfrac{3}{7} = $ —

(26) $\dfrac{4}{10} - \dfrac{2}{10} = $ —

(27) $\dfrac{3}{10} - \dfrac{1}{10} = $ —

(28) $\dfrac{4}{12} - \dfrac{2}{12} = $ —

(29) $\dfrac{7}{9} - \dfrac{1}{9} = $ —

(30) $\dfrac{4}{6} - \dfrac{2}{6} = $ —

(31) $\dfrac{2}{4} - \dfrac{1}{4} = $ —

(32) $\dfrac{5}{3} - \dfrac{2}{3} = $ —

(33) $\dfrac{3}{7} - \dfrac{1}{7} = $ —

(34) $\dfrac{4}{9} - \dfrac{3}{9} = $ —

(35) $\dfrac{2}{3} - \dfrac{1}{3} = $ —

(36) $\dfrac{8}{9} - \dfrac{2}{9} = $ —

(37) $\dfrac{6}{10} - \dfrac{2}{10} = $ —

(38) $\dfrac{2}{5} - \dfrac{1}{5} = $ —

(39) $\dfrac{3}{12} - \dfrac{1}{12} = $ —

(40) $\dfrac{6}{7} - \dfrac{1}{7} = $ —

(41) $\dfrac{5}{6} - \dfrac{1}{6} = $ —

(42) $\dfrac{5}{10} - \dfrac{3}{10} = $ —

(43) $\dfrac{4}{8} - \dfrac{3}{8} = $ —

(44) $\dfrac{3}{4} - \dfrac{1}{4} = $ —

(45) $\dfrac{6}{10} - \dfrac{2}{10} = $ —

Subtracting fractions with like denominators:

(1) $\dfrac{4}{10} - \dfrac{2}{10} = $ —

(2) $\dfrac{5}{7} - \dfrac{2}{7} = $ —

(3) $\dfrac{2}{8} - \dfrac{1}{8} = $ —

(4) $\dfrac{2}{6} - \dfrac{1}{6} = $ —

(5) $\dfrac{4}{9} - \dfrac{1}{9} = $ —

(6) $\dfrac{8}{8} - \dfrac{4}{8} = $ —

(7) $\dfrac{6}{10} - \dfrac{3}{10} = $ —

(8) $\dfrac{3}{8} - \dfrac{1}{8} = $ —

(9) $\dfrac{4}{9} - \dfrac{3}{9} = $ —

(10) $\dfrac{4}{7} - \dfrac{2}{7} = $ —

(11) $\dfrac{3}{10} - \dfrac{2}{10} = $ —

(12) $\dfrac{3}{3} - \dfrac{1}{3} = $ —

(13) $\dfrac{4}{8} - \dfrac{1}{8} = $ —

(14) $\dfrac{5}{12} - \dfrac{2}{12} = $ —

(15) $\dfrac{3}{4} - \dfrac{2}{4} = $ —

(16) $\dfrac{4}{6} - \dfrac{2}{6} = $ —

(17) $\dfrac{4}{10} - \dfrac{2}{10} = $ —

(18) $\dfrac{4}{5} - \dfrac{2}{5} = $ —

(19) $\dfrac{2}{12} - \dfrac{1}{12} = $ —

(20) $\dfrac{4}{9} - \dfrac{1}{9} = $ —

(21) $\dfrac{5}{7} - \dfrac{3}{7} = $ —

(22) $\dfrac{9}{4} - \dfrac{3}{4} = $ —

(23) $\dfrac{6}{8} - \dfrac{2}{8} = $ —

(24) $\dfrac{4}{6} - \dfrac{1}{6} = $ —

(25) $\dfrac{5}{7} - \dfrac{2}{7} = $ —

(26) $\dfrac{3}{10} - \dfrac{1}{10} = $ —

(27) $\dfrac{5}{7} - \dfrac{2}{7} = $ —

(28) $\dfrac{6}{5} - \dfrac{3}{5} = $ —

(29) $\dfrac{3}{6} - \dfrac{2}{6} = $ —

(30) $\dfrac{5}{12} - \dfrac{4}{12} = $ —

(31) $\dfrac{7}{9} - \dfrac{2}{9} = $ —

(32) $\dfrac{3}{7} - \dfrac{1}{7} = $ —

(33) $\dfrac{4}{4} - \dfrac{3}{4} = $ —

(34) $\dfrac{2}{10} - \dfrac{1}{10} = $ —

(35) $\dfrac{7}{10} - \dfrac{3}{10} = $ —

(36) $\dfrac{6}{10} - \dfrac{1}{10} = $ —

(37) $\dfrac{9}{7} - \dfrac{5}{7} = $ —

(38) $\dfrac{3}{5} - \dfrac{2}{5} = $ —

(39) $\dfrac{5}{8} - \dfrac{2}{8} = $ —

(40) $\dfrac{5}{6} - \dfrac{3}{6} = $ —

(41) $\dfrac{7}{12} - \dfrac{3}{12} = $ —

(42) $\dfrac{5}{9} - \dfrac{1}{9} = $ —

(43) $\dfrac{3}{8} - \dfrac{2}{8} = $ —

(44) $\dfrac{4}{7} - \dfrac{3}{7} = $ —

(45) $\dfrac{2}{5} - \dfrac{1}{5} = $ —

Subtracting fractions with like denominators:

(1) $\dfrac{2}{9} - \dfrac{1}{9} =$ —

(2) $\dfrac{6}{10} - \dfrac{5}{10} =$ —

(3) $\dfrac{5}{10} - \dfrac{3}{10} =$ —

(4) $\dfrac{5}{10} - \dfrac{4}{10} =$ —

(5) $\dfrac{6}{8} - \dfrac{1}{8} =$ —

(6) $\dfrac{3}{3} - \dfrac{2}{3} =$ —

(7) $\dfrac{3}{5} - \dfrac{2}{5} =$ —

(8) $\dfrac{5}{9} - \dfrac{2}{9} =$ —

(9) $\dfrac{4}{7} - \dfrac{1}{7} =$ —

(10) $\dfrac{2}{8} - \dfrac{1}{8} =$ —

(11) $\dfrac{5}{7} - \dfrac{2}{7} =$ —

(12) $\dfrac{2}{9} - \dfrac{1}{9} =$ —

(13) $\dfrac{5}{10} - \dfrac{4}{10} =$ —

(14) $\dfrac{4}{5} - \dfrac{1}{5} =$ —

(15) $\dfrac{3}{6} - \dfrac{2}{6} =$ —

(16) $\dfrac{4}{10} - \dfrac{2}{10} =$ —

(17) $\dfrac{4}{8} - \dfrac{3}{8} =$ —

(18) $\dfrac{4}{12} - \dfrac{2}{12} =$ —

(19) $\dfrac{2}{4} - \dfrac{1}{4} =$ —

(20) $\dfrac{7}{10} - \dfrac{4}{10} =$ —

(21) $\dfrac{6}{10} - \dfrac{4}{10} =$ —

(22) $\dfrac{5}{8} - \dfrac{3}{8} =$ —

(23) $\dfrac{3}{10} - \dfrac{1}{10} =$ —

(24) $\dfrac{3}{7} - \dfrac{2}{7} =$ —

(25) $\dfrac{3}{6} - \dfrac{2}{6} =$ —

(26) $\dfrac{5}{6} - \dfrac{2}{6} =$ —

(27) $\dfrac{4}{6} - \dfrac{2}{6} =$ —

(28) $\dfrac{2}{7} - \dfrac{1}{7} =$ —

(29) $\dfrac{5}{9} - \dfrac{4}{9} =$ —

(30) $\dfrac{3}{3} - \dfrac{1}{3} =$ —

(31) $\dfrac{5}{9} - \dfrac{4}{9} =$ —

(32) $\dfrac{4}{5} - \dfrac{1}{5} =$ —

(33) $\dfrac{4}{8} - \dfrac{3}{8} =$ —

(34) $\dfrac{5}{10} - \dfrac{3}{10} =$ —

(35) $\dfrac{4}{10} - \dfrac{3}{10} =$ —

(36) $\dfrac{4}{5} - \dfrac{2}{5} =$ —

(37) $\dfrac{4}{7} - \dfrac{2}{7} =$ —

(38) $\dfrac{3}{10} - \dfrac{2}{10} =$ —

(39) $\dfrac{4}{2} - \dfrac{1}{2} =$ —

(40) $\dfrac{2}{6} - \dfrac{1}{6} =$ —

(41) $\dfrac{3}{7} - \dfrac{2}{7} =$ —

(42) $\dfrac{5}{7} - \dfrac{4}{7} =$ —

(43) $\dfrac{5}{12} - \dfrac{4}{12} =$ —

(44) $\dfrac{3}{12} - \dfrac{1}{12} =$ —

(45) $\dfrac{3}{10} - \dfrac{2}{10} =$ —

Subtracting fractions with like denominators:

(1) $\dfrac{3}{10} - \dfrac{2}{10} = $ ___

(2) $\dfrac{4}{10} - \dfrac{3}{10} = $ ___

(3) $\dfrac{3}{7} - \dfrac{1}{7} = $ ___

(4) $\dfrac{2}{10} - \dfrac{1}{10} = $ ___

(5) $\dfrac{5}{8} - \dfrac{1}{8} = $ ___

(6) $\dfrac{5}{6} - \dfrac{2}{6} = $ ___

(7) $\dfrac{4}{7} - \dfrac{3}{7} = $ ___

(8) $\dfrac{5}{12} - \dfrac{2}{12} = $ ___

(9) $\dfrac{4}{5} - \dfrac{3}{5} = $ ___

(10) $\dfrac{3}{3} - \dfrac{2}{3} = $ ___

(11) $\dfrac{2}{4} - \dfrac{1}{4} = $ ___

(12) $\dfrac{3}{8} - \dfrac{1}{8} = $ ___

(13) $\dfrac{2}{12} - \dfrac{1}{12} = $ ___

(14) $\dfrac{2}{10} - \dfrac{1}{10} = $ ___

(15) $\dfrac{5}{12} - \dfrac{4}{12} = $ ___

(16) $\dfrac{5}{8} - \dfrac{4}{8} = $ ___

(17) $\dfrac{3}{6} - \dfrac{2}{6} = $ ___

(18) $\dfrac{3}{6} - \dfrac{2}{6} = $ ___

(19) $\dfrac{3}{4} - \dfrac{2}{4} = $ ___

(20) $\dfrac{2}{3} - \dfrac{1}{3} = $ ___

(21) $\dfrac{3}{5} - \dfrac{1}{5} = $ ___

(22) $\dfrac{2}{5} - \dfrac{1}{5} = $ ___

(23) $\dfrac{3}{4} - \dfrac{1}{4} = $ ___

(24) $\dfrac{5}{10} - \dfrac{3}{10} = $ ___

(25) $\dfrac{5}{10} - \dfrac{4}{10} = $ ___

(26) $\dfrac{6}{8} - \dfrac{4}{8} = $ ___

(27) $\dfrac{3}{8} - \dfrac{2}{8} = $ ___

(28) $\dfrac{3}{9} - \dfrac{2}{9} = $ ___

(29) $\dfrac{5}{9} - \dfrac{3}{9} = $ ___

(30) $\dfrac{2}{6} - \dfrac{1}{6} = $ ___

(31) $\dfrac{2}{2} - \dfrac{1}{2} = $ ___

(32) $\dfrac{3}{5} - \dfrac{2}{5} = $ ___

(33) $\dfrac{5}{9} - \dfrac{4}{9} = $ ___

(34) $\dfrac{4}{6} - \dfrac{3}{6} = $ ___

(35) $\dfrac{4}{6} - \dfrac{1}{6} = $ ___

(36) $\dfrac{3}{10} - \dfrac{2}{10} = $ ___

(37) $\dfrac{3}{12} - \dfrac{2}{12} = $ ___

(38) $\dfrac{6}{10} - \dfrac{2}{10} = $ ___

(39) $\dfrac{2}{12} - \dfrac{1}{12} = $ ___

(40) $\dfrac{3}{10} - \dfrac{1}{10} = $ ___

(41) $\dfrac{2}{12} - \dfrac{1}{12} = $ ___

(42) $\dfrac{6}{9} - \dfrac{3}{9} = $ ___

(43) $\dfrac{7}{6} - \dfrac{4}{6} = $ ___

(44) $\dfrac{4}{6} - \dfrac{2}{6} = $ ___

(45) $\dfrac{5}{4} - \dfrac{2}{4} = $ ___

Subtracting fractions with like denominators:

(1) $\dfrac{7}{7} - \dfrac{2}{7} = $ —

(2) $\dfrac{7}{8} - \dfrac{1}{8} = $ —

(3) $\dfrac{2}{12} - \dfrac{1}{12} = $ —

(4) $\dfrac{2}{3} - \dfrac{1}{3} = $ —

(5) $\dfrac{5}{7} - \dfrac{4}{7} = $ —

(6) $\dfrac{5}{8} - \dfrac{4}{8} = $ —

(7) $\dfrac{5}{9} - \dfrac{3}{9} = $ —

(8) $\dfrac{6}{12} - \dfrac{4}{12} = $ —

(9) $\dfrac{3}{9} - \dfrac{2}{9} = $ —

(10) $\dfrac{4}{5} - \dfrac{2}{5} = $ —

(11) $\dfrac{3}{4} - \dfrac{2}{4} = $ —

(12) $\dfrac{2}{10} - \dfrac{1}{10} = $ —

(13) $\dfrac{3}{10} - \dfrac{1}{10} = $ —

(14) $\dfrac{6}{9} - \dfrac{3}{9} = $ —

(15) $\dfrac{5}{12} - \dfrac{4}{12} = $ —

(16) $\dfrac{8}{7} - \dfrac{4}{7} = $ —

(17) $\dfrac{5}{10} - \dfrac{3}{10} = $ —

(18) $\dfrac{4}{7} - \dfrac{3}{7} = $ —

(19) $\dfrac{4}{8} - \dfrac{2}{8} = $ —

(20) $\dfrac{5}{3} - \dfrac{1}{3} = $ —

(21) $\dfrac{5}{5} - \dfrac{2}{5} = $ —

(22) $\dfrac{3}{6} - \dfrac{1}{6} = $ —

(23) $\dfrac{4}{8} - \dfrac{2}{8} = $ —

(24) $\dfrac{3}{4} - \dfrac{1}{4} = $ —

(25) $\dfrac{5}{5} - \dfrac{3}{5} = $ —

(26) $\dfrac{2}{9} - \dfrac{1}{9} = $ —

(27) $\dfrac{4}{6} - \dfrac{3}{6} = $ —

(28) $\dfrac{5}{10} - \dfrac{2}{10} = $ —

(29) $\dfrac{5}{10} - \dfrac{2}{10} = $ —

(30) $\dfrac{4}{7} - \dfrac{2}{7} = $ —

(31) $\dfrac{3}{4} - \dfrac{1}{4} = $ —

(32) $\dfrac{6}{10} - \dfrac{4}{10} = $ —

(33) $\dfrac{3}{9} - \dfrac{1}{9} = $ —

(34) $\dfrac{6}{9} - \dfrac{4}{9} = $ —

(35) $\dfrac{4}{2} - \dfrac{1}{2} = $ —

(36) $\dfrac{4}{10} - \dfrac{3}{10} = $ —

(37) $\dfrac{4}{6} - \dfrac{2}{6} = $ —

(38) $\dfrac{4}{6} - \dfrac{2}{6} = $ —

(39) $\dfrac{3}{12} - \dfrac{2}{12} = $ —

(40) $\dfrac{3}{9} - \dfrac{1}{9} = $ —

(41) $\dfrac{2}{8} - \dfrac{1}{8} = $ —

(42) $\dfrac{2}{7} - \dfrac{1}{7} = $ —

(43) $\dfrac{5}{7} - \dfrac{3}{7} = $ —

(44) $\dfrac{3}{4} - \dfrac{1}{4} = $ —

(45) $\dfrac{5}{12} - \dfrac{4}{12} = $ —

Subtracting fractions with like denominators:

(1) $\dfrac{4}{4} - \dfrac{2}{4} = $ ___

(2) $\dfrac{3}{10} - \dfrac{2}{10} = $ ___

(3) $\dfrac{4}{6} - \dfrac{2}{6} = $ ___

(4) $\dfrac{3}{12} - \dfrac{1}{12} = $ ___

(5) $\dfrac{5}{9} - \dfrac{4}{9} = $ ___

(6) $\dfrac{2}{3} - \dfrac{1}{3} = $ ___

(7) $\dfrac{7}{8} - \dfrac{4}{8} = $ ___

(8) $\dfrac{3}{7} - \dfrac{1}{7} = $ ___

(9) $\dfrac{5}{8} - \dfrac{3}{8} = $ ___

(10) $\dfrac{4}{10} - \dfrac{2}{10} = $ ___

(11) $\dfrac{4}{12} - \dfrac{3}{12} = $ ___

(12) $\dfrac{3}{4} - \dfrac{2}{4} = $ ___

(13) $\dfrac{3}{10} - \dfrac{1}{10} = $ ___

(14) $\dfrac{5}{8} - \dfrac{4}{8} = $ ___

(15) $\dfrac{6}{5} - \dfrac{1}{5} = $ ___

(16) $\dfrac{4}{8} - \dfrac{3}{8} = $ ___

(17) $\dfrac{2}{5} - \dfrac{1}{5} = $ ___

(18) $\dfrac{5}{9} - \dfrac{4}{9} = $ ___

(19) $\dfrac{5}{5} - \dfrac{2}{5} = $ ___

(20) $\dfrac{5}{6} - \dfrac{2}{6} = $ ___

(21) $\dfrac{4}{10} - \dfrac{2}{10} = $ ___

(22) $\dfrac{3}{7} - \dfrac{1}{7} = $ ___

(23) $\dfrac{7}{3} - \dfrac{1}{3} = $ ___

(24) $\dfrac{3}{12} - \dfrac{1}{12} = $ ___

(25) $\dfrac{4}{10} - \dfrac{3}{10} = $ ___

(26) $\dfrac{5}{7} - \dfrac{3}{7} = $ ___

(27) $\dfrac{6}{7} - \dfrac{3}{7} = $ ___

(28) $\dfrac{4}{9} - \dfrac{2}{9} = $ ___

(29) $\dfrac{3}{9} - \dfrac{1}{9} = $ ___

(30) $\dfrac{4}{8} - \dfrac{2}{8} = $ ___

(31) $\dfrac{3}{3} - \dfrac{2}{3} = $ ___

(32) $\dfrac{2}{7} - \dfrac{1}{7} = $ ___

(33) $\dfrac{3}{6} - \dfrac{1}{6} = $ ___

(34) $\dfrac{4}{9} - \dfrac{2}{9} = $ ___

(35) $\dfrac{4}{8} - \dfrac{3}{8} = $ ___

(36) $\dfrac{5}{12} - \dfrac{4}{12} = $ ___

(37) $\dfrac{9}{10} - \dfrac{1}{10} = $ ___

(38) $\dfrac{3}{12} - \dfrac{2}{12} = $ ___

(39) $\dfrac{3}{7} - \dfrac{2}{7} = $ ___

(40) $\dfrac{5}{7} - \dfrac{3}{7} = $ ___

(41) $\dfrac{2}{6} - \dfrac{1}{6} = $ ___

(42) $\dfrac{5}{10} - \dfrac{4}{10} = $ ___

(43) $\dfrac{4}{6} - \dfrac{2}{6} = $ ___

(44) $\dfrac{4}{10} - \dfrac{2}{10} = $ ___

(45) $\dfrac{10}{12} - \dfrac{1}{12} = $ ___

Subtracting fractions with like denominators:

(1) $\dfrac{2}{5} - \dfrac{1}{5} = $ ___

(2) $\dfrac{8}{8} - \dfrac{1}{8} = $ ___

(3) $\dfrac{5}{9} - \dfrac{3}{9} = $ ___

(4) $\dfrac{5}{8} - \dfrac{4}{8} = $ ___

(5) $\dfrac{5}{12} - \dfrac{4}{12} = $ ___

(6) $\dfrac{3}{5} - \dfrac{2}{5} = $ ___

(7) $\dfrac{5}{4} - \dfrac{2}{4} = $ ___

(8) $\dfrac{4}{9} - \dfrac{1}{9} = $ ___

(9) $\dfrac{2}{8} - \dfrac{1}{8} = $ ___

(10) $\dfrac{3}{12} - \dfrac{1}{12} = $ ___

(11) $\dfrac{2}{4} - \dfrac{1}{4} = $ ___

(12) $\dfrac{9}{7} - \dfrac{4}{7} = $ ___

(13) $\dfrac{5}{10} - \dfrac{2}{10} = $ ___

(14) $\dfrac{3}{6} - \dfrac{2}{6} = $ ___

(15) $\dfrac{6}{12} - \dfrac{3}{12} = $ ___

(16) $\dfrac{7}{8} - \dfrac{3}{8} = $ ___

(17) $\dfrac{4}{7} - \dfrac{3}{7} = $ ___

(18) $\dfrac{4}{3} - \dfrac{2}{3} = $ ___

(19) $\dfrac{9}{7} - \dfrac{1}{7} = $ ___

(20) $\dfrac{6}{10} - \dfrac{1}{10} = $ ___

(21) $\dfrac{4}{4} - \dfrac{1}{4} = $ ___

(22) $\dfrac{5}{5} - \dfrac{4}{5} = $ ___

(23) $\dfrac{3}{8} - \dfrac{2}{8} = $ ___

(24) $\dfrac{5}{5} - \dfrac{2}{5} = $ ___

(25) $\dfrac{4}{10} - \dfrac{2}{10} = $ ___

(26) $\dfrac{4}{5} - \dfrac{2}{5} = $ ___

(27) $\dfrac{4}{7} - \dfrac{3}{7} = $ ___

(28) $\dfrac{6}{6} - \dfrac{1}{6} = $ ___

(29) $\dfrac{9}{7} - \dfrac{1}{7} = $ ___

(30) $\dfrac{8}{12} - \dfrac{4}{12} = $ ___

(31) $\dfrac{8}{9} - \dfrac{3}{9} = $ ___

(32) $\dfrac{2}{9} - \dfrac{1}{9} = $ ___

(33) $\dfrac{7}{10} - \dfrac{1}{10} = $ ___

(34) $\dfrac{3}{7} - \dfrac{2}{7} = $ ___

(35) $\dfrac{4}{12} - \dfrac{3}{12} = $ ___

(36) $\dfrac{8}{9} - \dfrac{5}{9} = $ ___

(37) $\dfrac{5}{10} - \dfrac{1}{10} = $ ___

(38) $\dfrac{5}{12} - \dfrac{4}{12} = $ ___

(39) $\dfrac{5}{4} - \dfrac{2}{4} = $ ___

(40) $\dfrac{5}{6} - \dfrac{4}{6} = $ ___

(41) $\dfrac{3}{12} - \dfrac{2}{12} = $ ___

(42) $\dfrac{8}{8} - \dfrac{3}{8} = $ ___

(43) $\dfrac{4}{12} - \dfrac{2}{12} = $ ___

(44) $\dfrac{2}{12} - \dfrac{1}{12} = $ ___

(45) $\dfrac{6}{12} - \dfrac{4}{12} = $ ___

Subtracting fractions with like denominators:

(1) $\dfrac{9}{8} - \dfrac{1}{8} = \underline{\quad}$

(2) $\dfrac{5}{9} - \dfrac{1}{9} = \underline{\quad}$

(3) $\dfrac{4}{7} - \dfrac{1}{7} = \underline{\quad}$

(4) $\dfrac{4}{6} - \dfrac{3}{6} = \underline{\quad}$

(5) $\dfrac{3}{5} - \dfrac{2}{5} = \underline{\quad}$

(6) $\dfrac{8}{12} - \dfrac{5}{12} = \underline{\quad}$

(7) $\dfrac{5}{8} - \dfrac{2}{8} = \underline{\quad}$

(8) $\dfrac{2}{10} - \dfrac{1}{10} = \underline{\quad}$

(9) $\dfrac{5}{9} - \dfrac{2}{9} = \underline{\quad}$

(10) $\dfrac{6}{4} - \dfrac{1}{4} = \underline{\quad}$

(11) $\dfrac{4}{4} - \dfrac{3}{4} = \underline{\quad}$

(12) $\dfrac{4}{10} - \dfrac{3}{10} = \underline{\quad}$

(13) $\dfrac{8}{10} - \dfrac{4}{10} = \underline{\quad}$

(14) $\dfrac{7}{12} - \dfrac{1}{12} = \underline{\quad}$

(15) $\dfrac{6}{13} - \dfrac{4}{13} = \underline{\quad}$

(16) $\dfrac{4}{9} - \dfrac{2}{9} = \underline{\quad}$

(17) $\dfrac{3}{3} - \dfrac{2}{3} = \underline{\quad}$

(18) $\dfrac{5}{8} - \dfrac{1}{8} = \underline{\quad}$

(19) $\dfrac{6}{3} - \dfrac{1}{3} = \underline{\quad}$

(20) $\dfrac{4}{8} - \dfrac{2}{8} = \underline{\quad}$

(21) $\dfrac{7}{14} - \dfrac{5}{14} = \underline{\quad}$

(22) $\dfrac{9}{10} - \dfrac{3}{10} = \underline{\quad}$

(23) $\dfrac{5}{10} - \dfrac{4}{10} = \underline{\quad}$

(24) $\dfrac{5}{7} - \dfrac{2}{7} = \underline{\quad}$

(25) $\dfrac{4}{5} - \dfrac{2}{5} = \underline{\quad}$

(26) $\dfrac{2}{12} - \dfrac{1}{12} = \underline{\quad}$

(27) $\dfrac{7}{9} - \dfrac{3}{9} = \underline{\quad}$

(28) $\dfrac{7}{9} - \dfrac{1}{9} = \underline{\quad}$

(29) $\dfrac{8}{6} - \dfrac{1}{6} = \underline{\quad}$

(30) $\dfrac{9}{10} - \dfrac{4}{10} = \underline{\quad}$

(31) $\dfrac{8}{10} - \dfrac{4}{10} = \underline{\quad}$

(32) $\dfrac{3}{7} - \dfrac{2}{7} = \underline{\quad}$

(33) $\dfrac{7}{13} - \dfrac{1}{13} = \underline{\quad}$

(34) $\dfrac{5}{6} - \dfrac{2}{6} = \underline{\quad}$

(35) $\dfrac{4}{6} - \dfrac{3}{6} = \underline{\quad}$

(36) $\dfrac{6}{12} - \dfrac{5}{12} = \underline{\quad}$

(37) $\dfrac{9}{2} - \dfrac{1}{2} = \underline{\quad}$

(38) $\dfrac{2}{3} - \dfrac{1}{3} = \underline{\quad}$

(39) $\dfrac{8}{6} - \dfrac{2}{6} = \underline{\quad}$

(40) $\dfrac{5}{5} - \dfrac{3}{5} = \underline{\quad}$

(41) $\dfrac{5}{12} - \dfrac{1}{12} = \underline{\quad}$

(42) $\dfrac{5}{12} - \dfrac{3}{12} = \underline{\quad}$

(43) $\dfrac{5}{3} - \dfrac{2}{3} = \underline{\quad}$

(44) $\dfrac{3}{4} - \dfrac{2}{4} = \underline{\quad}$

(45) $\dfrac{8}{8} - \dfrac{4}{8} = \underline{\quad}$

Subtracting fractions with like denominators:

(1) $\dfrac{8}{12} - \dfrac{1}{12} =$ —

(2) $\dfrac{6}{8} - \dfrac{5}{8} =$ —

(3) $\dfrac{6}{5} - \dfrac{1}{5} =$ —

(4) $\dfrac{6}{9} - \dfrac{4}{9} =$ —

(5) $\dfrac{3}{7} - \dfrac{1}{7} =$ —

(6) $\dfrac{7}{10} - \dfrac{5}{10} =$ —

(7) $\dfrac{9}{10} - \dfrac{2}{10} =$ —

(8) $\dfrac{5}{8} - \dfrac{3}{8} =$ —

(9) $\dfrac{5}{12} - \dfrac{2}{12} =$ —

(10) $\dfrac{7}{10} - \dfrac{1}{10} =$ —

(11) $\dfrac{3}{12} - \dfrac{2}{12} =$ —

(12) $\dfrac{8}{13} - \dfrac{3}{13} =$ —

(13) $\dfrac{8}{7} - \dfrac{3}{7} =$ —

(14) $\dfrac{2}{6} - \dfrac{1}{6} =$ —

(15) $\dfrac{5}{9} - \dfrac{4}{9} =$ —

(16) $\dfrac{6}{8} - \dfrac{2}{8} =$ —

(17) $\dfrac{7}{9} - \dfrac{2}{9} =$ —

(18) $\dfrac{7}{12} - \dfrac{1}{12} =$ —

(19) $\dfrac{5}{6} - \dfrac{1}{6} =$ —

(20) $\dfrac{4}{12} - \dfrac{1}{12} =$ —

(21) $\dfrac{6}{8} - \dfrac{5}{8} =$ —

(22) $\dfrac{8}{12} - \dfrac{4}{12} =$ —

(23) $\dfrac{7}{8} - \dfrac{1}{8} =$ —

(24) $\dfrac{7}{8} - \dfrac{2}{8} =$ —

(25) $\dfrac{5}{7} - \dfrac{2}{7} =$ —

(26) $\dfrac{3}{6} - \dfrac{2}{6} =$ —

(27) $\dfrac{5}{12} - \dfrac{3}{12} =$ —

(28) $\dfrac{9}{5} - \dfrac{1}{5} =$ —

(29) $\dfrac{3}{3} - \dfrac{2}{3} =$ —

(30) $\dfrac{8}{15} - \dfrac{4}{15} =$ —

(31) $\dfrac{4}{10} - \dfrac{3}{10} =$ —

(32) $\dfrac{3}{4} - \dfrac{1}{4} =$ —

(33) $\dfrac{4}{12} - \dfrac{1}{12} =$ —

(34) $\dfrac{5}{4} - \dfrac{2}{4} =$ —

(35) $\dfrac{8}{9} - \dfrac{1}{9} =$ —

(36) $\dfrac{6}{13} - \dfrac{5}{13} =$ —

(37) $\dfrac{2}{10} - \dfrac{1}{10} =$ —

(38) $\dfrac{4}{10} - \dfrac{1}{10} =$ —

(39) $\dfrac{8}{7} - \dfrac{2}{7} =$ —

(40) $\dfrac{5}{7} - \dfrac{4}{7} =$ —

(41) $\dfrac{4}{2} - \dfrac{1}{2} =$ —

(42) $\dfrac{5}{10} - \dfrac{3}{10} =$ —

(43) $\dfrac{3}{12} - \dfrac{2}{12} =$ —

(44) $\dfrac{8}{8} - \dfrac{4}{8} =$ —

(45) $\dfrac{8}{13} - \dfrac{4}{13} =$ —

Subtracting fractions with like denominators:

(1) $\dfrac{8}{4} - \dfrac{1}{4} = $ —

(2) $\dfrac{4}{5} - \dfrac{1}{5} = $ —

(3) $\dfrac{6}{9} - \dfrac{1}{9} = $ —

(4) $\dfrac{4}{8} - \dfrac{3}{8} = $ —

(5) $\dfrac{4}{10} - \dfrac{1}{10} = $ —

(6) $\dfrac{9}{12} - \dfrac{5}{12} = $ —

(7) $\dfrac{4}{5} - \dfrac{2}{5} = $ —

(8) $\dfrac{3}{10} - \dfrac{2}{10} = $ —

(9) $\dfrac{8}{6} - \dfrac{2}{6} = $ —

(10) $\dfrac{7}{3} - \dfrac{1}{3} = $ —

(11) $\dfrac{2}{12} - \dfrac{1}{12} = $ —

(12) $\dfrac{4}{9} - \dfrac{3}{9} = $ —

(13) $\dfrac{4}{9} - \dfrac{3}{9} = $ —

(14) $\dfrac{5}{8} - \dfrac{1}{8} = $ —

(15) $\dfrac{9}{12} - \dfrac{4}{12} = $ —

(16) $\dfrac{3}{6} - \dfrac{2}{6} = $ —

(17) $\dfrac{3}{5} - \dfrac{1}{5} = $ —

(18) $\dfrac{7}{8} - \dfrac{1}{8} = $ —

(19) $\dfrac{2}{7} - \dfrac{1}{7} = $ —

(20) $\dfrac{5}{9} - \dfrac{3}{9} = $ —

(21) $\dfrac{6}{10} - \dfrac{5}{10} = $ —

(22) $\dfrac{5}{8} - \dfrac{4}{8} = $ —

(23) $\dfrac{5}{6} - \dfrac{1}{6} = $ —

(24) $\dfrac{5}{5} - \dfrac{2}{5} = $ —

(25) $\dfrac{3}{12} - \dfrac{2}{12} = $ —

(26) $\dfrac{5}{6} - \dfrac{2}{6} = $ —

(27) $\dfrac{8}{7} - \dfrac{3}{7} = $ —

(28) $\dfrac{2}{8} - \dfrac{1}{8} = $ —

(29) $\dfrac{5}{12} - \dfrac{1}{12} = $ —

(30) $\dfrac{5}{10} - \dfrac{4}{10} = $ —

(31) $\dfrac{3}{7} - \dfrac{2}{7} = $ —

(32) $\dfrac{5}{4} - \dfrac{1}{4} = $ —

(33) $\dfrac{9}{9} - \dfrac{1}{9} = $ —

(34) $\dfrac{4}{6} - \dfrac{2}{6} = $ —

(35) $\dfrac{4}{7} - \dfrac{1}{7} = $ —

(36) $\dfrac{7}{12} - \dfrac{5}{12} = $ —

(37) $\dfrac{3}{3} - \dfrac{1}{3} = $ —

(38) $\dfrac{4}{6} - \dfrac{2}{6} = $ —

(39) $\dfrac{6}{8} - \dfrac{2}{8} = $ —

(40) $\dfrac{4}{8} - \dfrac{3}{8} = $ —

(41) $\dfrac{7}{12} - \dfrac{3}{12} = $ —

(42) $\dfrac{5}{12} - \dfrac{3}{12} = $ —

(43) $\dfrac{4}{5} - \dfrac{2}{5} = $ —

(44) $\dfrac{3}{3} - \dfrac{1}{3} = $ —

(45) $\dfrac{8}{6} - \dfrac{4}{6} = $ —

Subtracting fractions with like denominators:

(1) $\dfrac{4}{2} - \dfrac{1}{2} = $ —

(2) $\dfrac{3}{5} - \dfrac{1}{5} = $ —

(3) $\dfrac{5}{7} - \dfrac{1}{7} = $ —

(4) $\dfrac{5}{7} - \dfrac{4}{7} = $ —

(5) $\dfrac{5}{5} - \dfrac{2}{5} = $ —

(6) $\dfrac{7}{10} - \dfrac{5}{10} = $ —

(7) $\dfrac{3}{8} - \dfrac{2}{8} = $ —

(8) $\dfrac{5}{10} - \dfrac{1}{10} = $ —

(9) $\dfrac{5}{5} - \dfrac{2}{5} = $ —

(10) $\dfrac{3}{7} - \dfrac{1}{7} = $ —

(11) $\dfrac{2}{9} - \dfrac{1}{9} = $ —

(12) $\dfrac{8}{8} - \dfrac{3}{8} = $ —

(13) $\dfrac{5}{6} - \dfrac{2}{6} = $ —

(14) $\dfrac{6}{7} - \dfrac{2}{7} = $ —

(15) $\dfrac{6}{12} - \dfrac{4}{12} = $ —

(16) $\dfrac{4}{5} - \dfrac{3}{5} = $ —

(17) $\dfrac{8}{12} - \dfrac{1}{12} = $ —

(18) $\dfrac{9}{12} - \dfrac{1}{12} = $ —

(19) $\dfrac{3}{8} - \dfrac{1}{8} = $ —

(20) $\dfrac{5}{7} - \dfrac{1}{7} = $ —

(21) $\dfrac{6}{7} - \dfrac{2}{7} = $ —

(22) $\dfrac{5}{12} - \dfrac{4}{12} = $ —

(23) $\dfrac{6}{8} - \dfrac{2}{8} = $ —

(24) $\dfrac{7}{5} - \dfrac{3}{5} = $ —

(25) $\dfrac{3}{6} - \dfrac{2}{6} = $ —

(26) $\dfrac{5}{3} - \dfrac{1}{3} = $ —

(27) $\dfrac{8}{9} - \dfrac{4}{9} = $ —

(28) $\dfrac{3}{5} - \dfrac{1}{5} = $ —

(29) $\dfrac{5}{3} - \dfrac{2}{3} = $ —

(30) $\dfrac{6}{6} - \dfrac{1}{6} = $ —

(31) $\dfrac{5}{8} - \dfrac{3}{8} = $ —

(32) $\dfrac{3}{6} - \dfrac{1}{6} = $ —

(33) $\dfrac{10}{12} - \dfrac{5}{12} = $ —

(34) $\dfrac{3}{8} - \dfrac{2}{8} = $ —

(35) $\dfrac{3}{10} - \dfrac{1}{10} = $ —

(36) $\dfrac{8}{3} - \dfrac{2}{3} = $ —

(37) $\dfrac{2}{6} - \dfrac{1}{6} = $ —

(38) $\dfrac{4}{8} - \dfrac{1}{8} = $ —

(39) $\dfrac{5}{4} - \dfrac{3}{4} = $ —

(40) $\dfrac{5}{9} - \dfrac{4}{9} = $ —

(41) $\dfrac{9}{4} - \dfrac{2}{4} = $ —

(42) $\dfrac{8}{5} - \dfrac{4}{5} = $ —

(43) $\dfrac{3}{10} - \dfrac{2}{10} = $ —

(44) $\dfrac{4}{4} - \dfrac{1}{4} = $ —

(45) $\dfrac{6}{12} - \dfrac{1}{12} = $ —

Subtracting fractions with like denominators:

(1) $\dfrac{2}{8} - \dfrac{1}{8} = \underline{}$

(2) $\dfrac{4}{12} - \dfrac{1}{12} = \underline{}$

(3) $\dfrac{9}{8} - \dfrac{5}{8} = \underline{}$

(4) $\dfrac{6}{9} - \dfrac{3}{9} = \underline{}$

(5) $\dfrac{6}{8} - \dfrac{3}{8} = \underline{}$

(6) $\dfrac{7}{9} - \dfrac{2}{9} = \underline{}$

(7) $\dfrac{5}{4} - \dfrac{2}{4} = \underline{}$

(8) $\dfrac{5}{9} - \dfrac{1}{9} = \underline{}$

(9) $\dfrac{5}{10} - \dfrac{3}{10} = \underline{}$

(10) $\dfrac{2}{12} - \dfrac{1}{12} = \underline{}$

(11) $\dfrac{3}{4} - \dfrac{2}{4} = \underline{}$

(12) $\dfrac{8}{7} - \dfrac{4}{7} = \underline{}$

(13) $\dfrac{5}{8} - \dfrac{4}{8} = \underline{}$

(14) $\dfrac{4}{9} - \dfrac{2}{9} = \underline{}$

(15) $\dfrac{6}{9} - \dfrac{1}{9} = \underline{}$

(16) $\dfrac{3}{9} - \dfrac{2}{9} = \underline{}$

(17) $\dfrac{5}{5} - \dfrac{1}{5} = \underline{}$

(18) $\dfrac{9}{15} - \dfrac{5}{15} = \underline{}$

(19) $\dfrac{2}{10} - \dfrac{1}{10} = \underline{}$

(20) $\dfrac{5}{12} - \dfrac{1}{12} = \underline{}$

(21) $\dfrac{7}{4} - \dfrac{2}{4} = \underline{}$

(22) $\dfrac{5}{8} - \dfrac{4}{8} = \underline{}$

(23) $\dfrac{7}{12} - \dfrac{2}{12} = \underline{}$

(24) $\dfrac{8}{7} - \dfrac{3}{7} = \underline{}$

(25) $\dfrac{4}{7} - \dfrac{3}{7} = \underline{}$

(26) $\dfrac{2}{12} - \dfrac{1}{12} = \underline{}$

(27) $\dfrac{6}{10} - \dfrac{4}{10} = \underline{}$

(28) $\dfrac{5}{5} - \dfrac{2}{5} = \underline{}$

(29) $\dfrac{4}{3} - \dfrac{1}{3} = \underline{}$

(30) $\dfrac{7}{12} - \dfrac{1}{12} = \underline{}$

(31) $\dfrac{3}{4} - \dfrac{1}{4} = \underline{}$

(32) $\dfrac{4}{3} - \dfrac{2}{3} = \underline{}$

(33) $\dfrac{6}{13} - \dfrac{5}{13} = \underline{}$

(34) $\dfrac{4}{6} - \dfrac{3}{6} = \underline{}$

(35) $\dfrac{2}{7} - \dfrac{1}{7} = \underline{}$

(36) $\dfrac{8}{6} - \dfrac{2}{6} = \underline{}$

(37) $\dfrac{4}{7} - \dfrac{2}{7} = \underline{}$

(38) $\dfrac{3}{5} - \dfrac{2}{5} = \underline{}$

(39) $\dfrac{6}{9} - \dfrac{3}{9} = \underline{}$

(40) $\dfrac{8}{9} - \dfrac{1}{9} = \underline{}$

(41) $\dfrac{3}{10} - \dfrac{1}{10} = \underline{}$

(42) $\dfrac{5}{12} - \dfrac{4}{12} = \underline{}$

(43) $\dfrac{4}{10} - \dfrac{3}{10} = \underline{}$

(44) $\dfrac{4}{6} - \dfrac{1}{6} = \underline{}$

(45) $\dfrac{9}{5} - \dfrac{1}{5} = \underline{}$

Subtracting fractions with like denominators:

(1) $\dfrac{3}{12} - \dfrac{2}{12} = $ ___

(2) $\dfrac{4}{6} - \dfrac{2}{6} = $ ___

(3) $\dfrac{7}{7} - \dfrac{5}{7} = $ ___

(4) $\dfrac{2}{7} - \dfrac{1}{7} = $ ___

(5) $\dfrac{2}{4} - \dfrac{1}{4} = $ ___

(6) $\dfrac{5}{8} - \dfrac{2}{8} = $ ___

(7) $\dfrac{5}{8} - \dfrac{4}{8} = $ ___

(8) $\dfrac{3}{9} - \dfrac{2}{9} = $ ___

(9) $\dfrac{7}{12} - \dfrac{3}{12} = $ ___

(10) $\dfrac{4}{6} - \dfrac{2}{6} = $ ___

(11) $\dfrac{3}{7} - \dfrac{1}{7} = $ ___

(12) $\dfrac{8}{6} - \dfrac{4}{6} = $ ___

(13) $\dfrac{2}{3} - \dfrac{1}{3} = $ ___

(14) $\dfrac{4}{10} - \dfrac{2}{10} = $ ___

(15) $\dfrac{6}{7} - \dfrac{1}{7} = $ ___

(16) $\dfrac{5}{8} - \dfrac{3}{8} = $ ___

(17) $\dfrac{4}{8} - \dfrac{1}{8} = $ ___

(18) $\dfrac{9}{10} - \dfrac{5}{10} = $ ___

(19) $\dfrac{3}{4} - \dfrac{2}{4} = $ ___

(20) $\dfrac{3}{8} - \dfrac{1}{8} = $ ___

(21) $\dfrac{7}{5} - \dfrac{2}{5} = $ ___

(22) $\dfrac{9}{5} - \dfrac{1}{5} = $ ___

(23) $\dfrac{6}{7} - \dfrac{1}{7} = $ ___

(24) $\dfrac{6}{8} - \dfrac{3}{8} = $ ___

(25) $\dfrac{6}{9} - \dfrac{4}{9} = $ ___

(26) $\dfrac{3}{12} - \dfrac{2}{12} = $ ___

(27) $\dfrac{9}{12} - \dfrac{4}{12} = $ ___

(28) $\dfrac{4}{10} - \dfrac{2}{10} = $ ___

(29) $\dfrac{3}{12} - \dfrac{1}{12} = $ ___

(30) $\dfrac{7}{4} - \dfrac{1}{4} = $ ___

(31) $\dfrac{3}{8} - \dfrac{1}{8} = $ ___

(32) $\dfrac{3}{9} - \dfrac{1}{9} = $ ___

(33) $\dfrac{6}{8} - \dfrac{5}{8} = $ ___

(34) $\dfrac{8}{7} - \dfrac{3}{7} = $ ___

(35) $\dfrac{5}{7} - \dfrac{2}{7} = $ ___

(36) $\dfrac{8}{7} - \dfrac{2}{7} = $ ___

(37) $\dfrac{4}{8} - \dfrac{2}{8} = $ ___

(38) $\dfrac{3}{12} - \dfrac{1}{12} = $ ___

(39) $\dfrac{5}{10} - \dfrac{3}{10} = $ ___

(40) $\dfrac{3}{6} - \dfrac{1}{6} = $ ___

(41) $\dfrac{3}{5} - \dfrac{1}{5} = $ ___

(42) $\dfrac{9}{13} - \dfrac{4}{13} = $ ___

(43) $\dfrac{5}{12} - \dfrac{4}{12} = $ ___

(44) $\dfrac{5}{8} - \dfrac{2}{8} = $ ___

(45) $\dfrac{6}{8} - \dfrac{1}{8} = $ ___

Subtracting fractions with like denominators:

(1) $\dfrac{7}{7} - \dfrac{2}{7} = \underline{}$

(2) $\dfrac{5}{6} - \dfrac{1}{6} = \underline{}$

(3) $\dfrac{6}{12} - \dfrac{5}{12} = \underline{}$

(4) $\dfrac{5}{10} - \dfrac{4}{10} = \underline{}$

(5) $\dfrac{5}{12} - \dfrac{2}{12} = \underline{}$

(6) $\dfrac{8}{9} - \dfrac{2}{9} = \underline{}$

(7) $\dfrac{6}{12} - \dfrac{1}{12} = \underline{}$

(8) $\dfrac{4}{10} - \dfrac{2}{10} = \underline{}$

(9) $\dfrac{7}{12} - \dfrac{3}{12} = \underline{}$

(10) $\dfrac{5}{9} - \dfrac{3}{9} = \underline{}$

(11) $\dfrac{6}{9} - \dfrac{3}{9} = \underline{}$

(12) $\dfrac{6}{8} - \dfrac{4}{8} = \underline{}$

(13) $\dfrac{3}{5} - \dfrac{2}{5} = \underline{}$

(14) $\dfrac{4}{7} - \dfrac{1}{7} = \underline{}$

(15) $\dfrac{8}{6} - \dfrac{1}{6} = \underline{}$

(16) $\dfrac{2}{8} - \dfrac{1}{8} = \underline{}$

(17) $\dfrac{5}{8} - \dfrac{3}{8} = \underline{}$

(18) $\dfrac{6}{9} - \dfrac{5}{9} = \underline{}$

(19) $\dfrac{7}{7} - \dfrac{4}{7} = \underline{}$

(20) $\dfrac{4}{6} - \dfrac{2}{6} = \underline{}$

(21) $\dfrac{5}{6} - \dfrac{2}{6} = \underline{}$

(22) $\dfrac{9}{8} - \dfrac{3}{8} = \underline{}$

(23) $\dfrac{4}{10} - \dfrac{1}{10} = \underline{}$

(24) $\dfrac{8}{7} - \dfrac{3}{7} = \underline{}$

(25) $\dfrac{4}{3} - \dfrac{2}{3} = \underline{}$

(26) $\dfrac{3}{5} - \dfrac{2}{5} = \underline{}$

(27) $\dfrac{6}{10} - \dfrac{4}{10} = \underline{}$

(28) $\dfrac{4}{4} - \dfrac{1}{4} = \underline{}$

(29) $\dfrac{4}{9} - \dfrac{1}{9} = \underline{}$

(30) $\dfrac{5}{9} - \dfrac{1}{9} = \underline{}$

(31) $\dfrac{6}{9} - \dfrac{3}{9} = \underline{}$

(32) $\dfrac{5}{12} - \dfrac{1}{12} = \underline{}$

(33) $\dfrac{7}{12} - \dfrac{5}{12} = \underline{}$

(34) $\dfrac{6}{8} - \dfrac{4}{8} = \underline{}$

(35) $\dfrac{3}{12} - \dfrac{2}{12} = \underline{}$

(36) $\dfrac{5}{8} - \dfrac{2}{8} = \underline{}$

(37) $\dfrac{4}{4} - \dfrac{1}{4} = \underline{}$

(38) $\dfrac{5}{9} - \dfrac{2}{9} = \underline{}$

(39) $\dfrac{7}{12} - \dfrac{3}{12} = \underline{}$

(40) $\dfrac{6}{7} - \dfrac{2}{7} = \underline{}$

(41) $\dfrac{2}{8} - \dfrac{1}{8} = \underline{}$

(42) $\dfrac{8}{6} - \dfrac{4}{6} = \underline{}$

(43) $\dfrac{3}{2} - \dfrac{1}{2} = \underline{}$

(44) $\dfrac{3}{6} - \dfrac{2}{6} = \underline{}$

(45) $\dfrac{5}{7} - \dfrac{1}{7} = \underline{}$

Subtracting fractions with like denominators:

(1) $\dfrac{5}{10} - \dfrac{2}{10} = $ ___

(2) $\dfrac{4}{12} - \dfrac{3}{12} = $ ___

(3) $\dfrac{7}{10} - \dfrac{5}{10} = $ ___

(4) $\dfrac{7}{8} - \dfrac{1}{8} = $ ___

(5) $\dfrac{4}{7} - \dfrac{2}{7} = $ ___

(6) $\dfrac{5}{5} - \dfrac{2}{5} = $ ___

(7) $\dfrac{5}{12} - \dfrac{3}{12} = $ ___

(8) $\dfrac{3}{10} - \dfrac{1}{10} = $ ___

(9) $\dfrac{4}{8} - \dfrac{3}{8} = $ ___

(10) $\dfrac{6}{3} - \dfrac{2}{3} = $ ___

(11) $\dfrac{4}{6} - \dfrac{1}{6} = $ ___

(12) $\dfrac{6}{12} - \dfrac{4}{12} = $ ___

(13) $\dfrac{3}{8} - \dfrac{2}{8} = $ ___

(14) $\dfrac{4}{8} - \dfrac{2}{8} = $ ___

(15) $\dfrac{4}{12} - \dfrac{1}{12} = $ ___

(16) $\dfrac{5}{6} - \dfrac{1}{6} = $ ___

(17) $\dfrac{4}{7} - \dfrac{1}{7} = $ ___

(18) $\dfrac{7}{13} - \dfrac{5}{13} = $ ___

(19) $\dfrac{5}{8} - \dfrac{1}{8} = $ ___

(20) $\dfrac{3}{12} - \dfrac{2}{12} = $ ___

(21) $\dfrac{5}{2} - \dfrac{1}{2} = $ ___

(22) $\dfrac{7}{9} - \dfrac{5}{9} = $ ___

(23) $\dfrac{4}{12} - \dfrac{1}{12} = $ ___

(24) $\dfrac{8}{4} - \dfrac{2}{4} = $ ___

(25) $\dfrac{3}{5} - \dfrac{2}{5} = $ ___

(26) $\dfrac{3}{6} - \dfrac{1}{6} = $ ___

(27) $\dfrac{8}{6} - \dfrac{3}{6} = $ ___

(28) $\dfrac{5}{6} - \dfrac{2}{6} = $ ___

(29) $\dfrac{3}{10} - \dfrac{2}{10} = $ ___

(30) $\dfrac{6}{8} - \dfrac{4}{8} = $ ___

(31) $\dfrac{2}{3} - \dfrac{1}{3} = $ ___

(32) $\dfrac{3}{9} - \dfrac{1}{9} = $ ___

(33) $\dfrac{10}{10} - \dfrac{5}{10} = $ ___

(34) $\dfrac{7}{7} - \dfrac{4}{7} = $ ___

(35) $\dfrac{4}{5} - \dfrac{1}{5} = $ ___

(36) $\dfrac{5}{3} - \dfrac{2}{3} = $ ___

(37) $\dfrac{3}{4} - \dfrac{1}{4} = $ ___

(38) $\dfrac{4}{12} - \dfrac{1}{12} = $ ___

(39) $\dfrac{6}{9} - \dfrac{3}{9} = $ ___

(40) $\dfrac{5}{8} - \dfrac{2}{8} = $ ___

(41) $\dfrac{3}{12} - \dfrac{2}{12} = $ ___

(42) $\dfrac{8}{10} - \dfrac{4}{10} = $ ___

(43) $\dfrac{4}{4} - \dfrac{2}{4} = $ ___

(44) $\dfrac{4}{7} - \dfrac{3}{7} = $ ___

(45) $\dfrac{5}{8} - \dfrac{1}{8} = $ ___

Subtracting fractions with like denominators:

(1) $\dfrac{4}{9} - \dfrac{1}{9} = \underline{}$

(2) $\dfrac{4}{9} - \dfrac{1}{9} = \underline{}$

(3) $\dfrac{7}{15} - \dfrac{5}{15} = \underline{}$

(4) $\dfrac{4}{10} - \dfrac{2}{10} = \underline{}$

(5) $\dfrac{5}{8} - \dfrac{1}{8} = \underline{}$

(6) $\dfrac{5}{5} - \dfrac{2}{5} = \underline{}$

(7) $\dfrac{4}{8} - \dfrac{3}{8} = \underline{}$

(8) $\dfrac{5}{6} - \dfrac{2}{6} = \underline{}$

(9) $\dfrac{6}{7} - \dfrac{3}{7} = \underline{}$

(10) $\dfrac{3}{6} - \dfrac{1}{6} = \underline{}$

(11) $\dfrac{3}{12} - \dfrac{1}{12} = \underline{}$

(12) $\dfrac{8}{9} - \dfrac{4}{9} = \underline{}$

(13) $\dfrac{4}{8} - \dfrac{2}{8} = \underline{}$

(14) $\dfrac{4}{8} - \dfrac{3}{8} = \underline{}$

(15) $\dfrac{6}{3} - \dfrac{1}{3} = \underline{}$

(16) $\dfrac{4}{6} - \dfrac{3}{6} = \underline{}$

(17) $\dfrac{3}{10} - \dfrac{1}{10} = \underline{}$

(18) $\dfrac{8}{8} - \dfrac{5}{8} = \underline{}$

(19) $\dfrac{5}{7} - \dfrac{1}{7} = \underline{}$

(20) $\dfrac{3}{7} - \dfrac{2}{7} = \underline{}$

(21) $\dfrac{8}{4} - \dfrac{2}{4} = \underline{}$

(22) $\dfrac{6}{5} - \dfrac{2}{5} = \underline{}$

(23) $\dfrac{4}{12} - \dfrac{1}{12} = \underline{}$

(24) $\dfrac{5}{5} - \dfrac{3}{5} = \underline{}$

(25) $\dfrac{4}{9} - \dfrac{2}{9} = \underline{}$

(26) $\dfrac{2}{5} - \dfrac{1}{5} = \underline{}$

(27) $\dfrac{7}{7} - \dfrac{4}{7} = \underline{}$

(28) $\dfrac{4}{5} - \dfrac{1}{5} = \underline{}$

(29) $\dfrac{4}{9} - \dfrac{3}{9} = \underline{}$

(30) $\dfrac{5}{10} - \dfrac{1}{10} = \underline{}$

(31) $\dfrac{5}{7} - \dfrac{2}{7} = \underline{}$

(32) $\dfrac{4}{8} - \dfrac{3}{8} = \underline{}$

(33) $\dfrac{7}{13} - \dfrac{5}{13} = \underline{}$

(34) $\dfrac{7}{10} - \dfrac{3}{10} = \underline{}$

(35) $\dfrac{4}{12} - \dfrac{2}{12} = \underline{}$

(36) $\dfrac{5}{6} - \dfrac{2}{6} = \underline{}$

(37) $\dfrac{4}{12} - \dfrac{2}{12} = \underline{}$

(38) $\dfrac{4}{7} - \dfrac{3}{7} = \underline{}$

(39) $\dfrac{5}{8} - \dfrac{3}{8} = \underline{}$

(40) $\dfrac{3}{10} - \dfrac{1}{10} = \underline{}$

(41) $\dfrac{2}{9} - \dfrac{1}{9} = \underline{}$

(42) $\dfrac{7}{12} - \dfrac{4}{12} = \underline{}$

(43) $\dfrac{4}{6} - \dfrac{2}{6} = \underline{}$

(44) $\dfrac{4}{5} - \dfrac{2}{5} = \underline{}$

(45) $\dfrac{8}{12} - \dfrac{1}{12} = \underline{}$

Finding Fractions from a model – Answer Key (10 Days)

#	DAY 1	DAY 3	DAY 5	DAY 7	DAY 9
(1)	$\frac{1}{3}$	$\frac{1}{12}$	$\frac{5}{8}$	$\frac{6}{12}$	$\frac{2}{4}$
(2)	$\frac{1}{4}$	$\frac{6}{11}$	$\frac{8}{11}$	$\frac{1}{5}$	$\frac{4}{8}$
(3)	$\frac{2}{5}$	$\frac{4}{10}$	$\frac{2}{4}$	$\frac{2}{9}$	$\frac{5}{7}$
(4)	$\frac{1}{2}$	$\frac{5}{6}$	$\frac{3}{9}$	$\frac{3}{4}$	$\frac{8}{12}$
(5)	$\frac{3}{6}$	$\frac{2}{6}$	$\frac{4}{10}$	$\frac{4}{12}$	$\frac{1}{10}$
(6)	$\frac{2}{7}$	$\frac{3}{12}$	$\frac{1}{4}$	$\frac{7}{12}$	$\frac{8}{9}$
(7)	$\frac{4}{6}$	$\frac{7}{11}$	$\frac{5}{7}$	$\frac{8}{10}$	$\frac{2}{11}$
(8)	$\frac{5}{8}$	$\frac{9}{10}$	$\frac{6}{11}$	$\frac{1}{6}$	$\frac{7}{10}$
(9)	$\frac{6}{10}$	$\frac{1}{8}$	$\frac{9}{12}$	$\frac{2}{8}$	$\frac{1}{6}$
(10)	$\frac{9}{11}$	$\frac{6}{9}$	$\frac{2}{5}$	$\frac{3}{7}$	$\frac{4}{9}$
(11)	$\frac{2}{4}$	$\frac{4}{11}$	$\frac{3}{11}$	$\frac{4}{11}$	$\frac{3}{11}$
(12)	$\frac{3}{6}$	$\frac{5}{10}$	$\frac{4}{5}$	$\frac{6}{11}$	$\frac{5}{10}$
(13)	$\frac{1}{6}$	$\frac{8}{9}$	$\frac{1}{12}$	$\frac{9}{11}$	$\frac{9}{12}$
(14)	$\frac{7}{9}$	$\frac{2}{7}$	$\frac{6}{9}$	$\frac{1}{4}$	$\frac{1}{8}$
(15)	$\frac{2}{10}$	$\frac{3}{9}$	$\frac{8}{9}$	$\frac{2}{6}$	$\frac{6}{11}$
(16)	$\frac{8}{10}$	$\frac{1}{10}$	$\frac{2}{7}$	$\frac{4}{7}$	$\frac{2}{5}$
(17)	$\frac{1}{5}$	$\frac{6}{7}$	$\frac{5}{10}$	$\frac{5}{10}$	$\frac{3}{6}$
(18)	$\frac{4}{7}$	$\frac{4}{8}$	$\frac{7}{9}$	$\frac{7}{9}$	$\frac{5}{11}$

#	DAY 2	DAY 4	DAY 6	DAY 8	DAY 10
(1)	$\frac{5}{11}$	$\frac{5}{12}$	$\frac{1}{11}$	$\frac{8}{11}$	$\frac{4}{6}$
(2)	$\frac{3}{5}$	$\frac{1}{7}$	$\frac{3}{10}$	$\frac{1}{7}$	$\frac{7}{11}$
(3)	$\frac{4}{9}$	$\frac{2}{8}$	$\frac{6}{7}$	$\frac{2}{5}$	$\frac{1}{11}$
(4)	$\frac{7}{12}$	$\frac{3}{8}$	$\frac{4}{9}$	$\frac{3}{11}$	$\frac{8}{11}$
(5)	$\frac{5}{9}$	$\frac{7}{10}$	$\frac{5}{6}$	$\frac{5}{8}$	$\frac{4}{12}$
(6)	$\frac{2}{3}$	$\frac{9}{11}$	$\frac{7}{11}$	$\frac{6}{10}$	$\frac{5}{6}$
(7)	$\frac{3}{11}$	$\frac{2}{12}$	$\frac{1}{10}$	$\frac{9}{12}$	$\frac{10}{11}$
(8)	$\frac{4}{5}$	$\frac{5}{11}$	$\frac{2}{11}$	$\frac{1}{9}$	$\frac{2}{12}$
(9)	$\frac{1}{4}$	$\frac{1}{3}$	$\frac{3}{7}$	$\frac{2}{7}$	$\frac{3}{10}$
(10)	$\frac{6}{8}$	$\frac{3}{5}$	$\frac{6}{8}$	$\frac{3}{6}$	$\frac{6}{12}$
(11)	$\frac{9}{12}$	$\frac{4}{7}$	$\frac{9}{10}$	$\frac{5}{11}$	$\frac{9}{11}$
(12)	$\frac{2}{5}$	$\frac{6}{10}$	$\frac{1}{8}$	$\frac{8}{9}$	$\frac{1}{4}$
(13)	$\frac{3}{10}$	$\frac{8}{10}$	$\frac{4}{11}$	$\frac{1}{3}$	$\frac{4}{10}$
(14)	$\frac{7}{8}$	$\frac{1}{9}$	$\frac{7}{8}$	$\frac{2}{12}$	$\frac{5}{12}$
(15)	$\frac{1}{11}$	$\frac{2}{6}$	$\frac{2}{3}$	$\frac{4}{6}$	$\frac{7}{8}$
(16)	$\frac{5}{7}$	$\frac{3}{6}$	$\frac{5}{9}$	$\frac{6}{9}$	$\frac{2}{6}$
(17)	$\frac{8}{11}$	$\frac{7}{12}$	$\frac{8}{12}$	$\frac{7}{10}$	$\frac{3}{7}$
(18)	$\frac{3}{4}$	$\frac{4}{6}$	$\frac{3}{12}$	$\frac{1}{12}$	$\frac{1}{3}$

Comparing Fractions - Answer Key (10 Days)

DAY 1

(1) >	(2) >	(3) <
(4) <	(5) >	(6) >
(7) >	(8) <	(9) <
(10) >	(11) >	(12) >
(13) <	(14) =	(15) <
(16) <	(17) >	(18) >
(19) <	(20) >	(21)
(22) >	(23) <	(24) <
(25) =	(26) >	(27) >
(28) <	(29) <	(30) >
(31) >	(32) >	(33) <
(34) <	(35) <	(36) <
(37) >	(38) =	(39) <
(40) >	(41) =	(42) <
(43)	(44) =	(45) <

DAY 2

(1) >	(2) >	(3) <
(4) <	(5) >	(6) <
(7) >	(8) <	(9) <
(10) <	(11) >	(12) >
(13) >	(14) >	(15) <
(16) >	(17) <	(18) <
(19) >	(20) <	(21) <
(22) >	(23) =	(24) <
(25) <	(26) >	(27) >
(28) >	(29) >	(30) <
(31) >	(32) <	(33) <
(34) >	(35) <	(36) <
(37) <	(38) >	(39) >
(40) <	(41) <	(42) <
(43) >	(44) <	(45) <

DAY 3

(1) >	(2) <	(3) <
(4) >	(5) =	(6) <
(7) >	(8) =	(9) >
(10) >	(11) =	(12) <
(13) <	(14) >	(15) >
(16) <	(17) <	(18) >
(19) >	(20) >	(21) <
(22) >	(23) =	(24) >
(25) >	(26) <	(27) <
(28) <	(29) <	(30) >
(31) >	(32) >	(33) >
(34) >	(35) <	(36) <
(37) >	(38) =	(39) <
(40) <	(41) <	(42) >
(43) <	(44) >	(45) >

DAY 4

(1) >	(2) =	(3) <
(4) >	(5) <	(6) <
(7) <	(8) =	(9) <
(10) <	(11) <	(12) >
(13) =	(14) <	(15) <
(16) <	(17) <	(18) >
(19) <	(20) >	(21) <
(22) <	(23) <	(24) >
(25) <	(26) =	(27) >
(28) <	(29) <	(30) >
(31) <	(32) =	(33) <
(34) <	(35) <	(36) <
(37) <	(38) =	(39) <
(40) >	(41) =	(42) >
(43) >	(44) >	(45) <

DAY 5

(1) >	(2) <	(3) =
(4) >	(5) <	(6) >
(7) >	(8) <	(9) >
(10) =	(11) >	(12) >
(13) >	(14) <	(15) <
(16) =	(17) >	(18) >
(19) =	(20) =	(21) >
(22) =	(23) <	(24) >
(25) <	(26) >	(27)
(28) =	(29) >	(30) <
(31) >	(32) <	(33) >
(34) =	(35) <	(36) =
(37) <	(38) =	(39) >
(40) <	(41) <	(42) <
(43) =	(44) >	(45) >

DAY 6

(1) <	(2) <	(3) >
(4) >	(5) <	(6) >
(7) =	(8) <	(9) >
(10) <	(11) =	(12) >
(13) =	(14) >	(15) =
(16) =	(17) <	(18) >
(19) =	(20) <	(21) >
(22) <	(23) <	(24) >
(25) >	(26) >	(27) >
(28) <	(29) <	(30) >
(31) <	(32) <	(33) =
(34) <	(35) <	(36) >
(37) <	(38) <	(39) >
(40) <	(41) <	(42) =
(43) =	(44) >	(45) >

DAY 7

(1) >	(2) <	(3) >
(4) <	(5) >	(6) >
(7) <	(8) <	(9) >
(10) >	(11) <	(12) >
(13) <	(14) <	(15) =
(16) <	(17) <	(18) <
(19) <	(20) <	(21) >
(22) <	(23) >	(24)
(25) <	(26) >	(27) =
(28) =	(29) <	(30)
(31) <	(32) <	(33) >
(34) <	(35) <	(36)
(37) =	(38) <	(39) >
(40) =	(41) <	(42) =
(43) =	(44) <	(45) >

DAY 8

(1) >	(2) <	(3) >
(4) <	(5) <	(6) >
(7) <	(8) <	(9) =
(10) >	(11) <	(12) >
(13) >	(14) <	(15) =
(16) <	(17) >	(18) <
(19) <	(20) <	(21) >
(22) =	(23) >	(24) <
(25) >	(26) <	(27) <
(28) >	(29) <	(30) <
(31) <	(32) >	(33) =
(34) >	(35) <	(36) <
(37) >	(38) <	(39)
(40) >	(41) <	(42) <
(43) <	(44) >	(45) <

DAY 9

(1) <	(2) <	(3) <
(4) =	(5) <	(6) >
(7) =	(8) >	(9) <
(10) =	(11) <	(12) =
(13) >	(14) <	(15) <
(16) <	(17) <	(18) <
(19) >	(20) >	(21) <
(22) =	(23) <	(24) <
(25) >	(26) >	(27) <
(28) <	(29) <	(30) >
(31) <	(32) <	(33) <
(34) >	(35) <	(36) <
(37) =	(38) <	(39) <
(40) <	(41) <	(42) <
(43) >	(44) >	(45) <

DAY 10

(1) =	(2) >	(3) <
(4) <	(5) <	(6) >
(7) =	(8) <	(9) =
(10) <	(11) >	(12) <
(13) <	(14) <	(15) <
(16) <	(17) >	(18) <
(19) >	(20) <	(21) <
(22) <	(23) >	(24) <
(25) =	(26) >	(27) =
(28) <	(29) >	(30) =
(31) =	(32) <	(33) <
(34) <	(35) >	(36) =
(37) =	(38) >	(39) <
(40) =	(41) <	(42) <
(43) >	(44) >	(45) <

Reduce Fractions – Answer Key (10 Days)

DAY 1

(1) 1/2	(2) 1/2	(3) 3/4
(4) 2/3	(5) 1/5	(6) 7/9
(7) 2/7	(8) 4/5	(9) 7/8
(10) 3/4	(11) 1/3	(12) 2/3
(13) 1/2	(14) 1/2	(15) 1/3
(16) 1/4	(17) 3/4	(18) 1/3
(19) 2/5	(20) 3/4	(21) 7/10
(22) 1/3	(23) 1/4	(24) 4/5
(25) 2/5	(26) 1/5	(27) 4/5
(28) 3/4	(29) 2/3	(30) 1/2
(31) 3/4	(32) 5/6	(33) 3/4
(34) 2/3	(35) 4/5	(36) 1/4
(37) 1/5	(38) 1/2	(39) 4/5
(40) 2/5	(41) 1/3	(42) 3/7
(43) 1/2	(44) 1/5	(45) 2/5

DAY 2

(1) 4/5	(2) 1/4	(3) 3/5
(4) 1/2	(5) 1/3	(6) 1/5
(7) 1/2	(8) 3/10	(9) 2/3
(10) 2/3	(11) 1/7	(12) 1/2
(13) 3/4	(14) 1/3	(15) 6/11
(16) 3/4	(17) 1/3	(18) 3/10
(19) 1/2	(20) 2/3	(21) 6/7
(22) 1/2	(23) 1/9	(24) 7/8
(25) 2/3	(26) 1/2	(27) 8/9
(28) 3/4	(29) 4/5	(30) 7/10
(31) 4/5	(32) 3/4	(33) 1/2
(34) 2/7	(35) 1/2	(36) 6/7
(37) 2/3	(38) 1/4	(39) 3/4
(40) 1/5	(41) 3/4	(42) 2/3
(43) 3/5	(44) 1/11	(45) 4/5

DAY 3

(1) 7/8	(2) 1/6	(3) 4/5
(4) 4/5	(5) 2/5	(6) 1/2
(7) 4/5	(8) 1/2	(9) 3/4
(10) 1/2	(11) 1/2	(12) 1/2
(13) 1/3	(14) 1/2	(15) 7/9
(16) 3/4	(17) 1/3	(18) 1/2
(19) 2/5	(20) 2/5	(21) 1/5
(22) 3/4	(23) 1/2	(24) 4/5
(25) 4/5	(26) 1/2	(27) 7/10
(28) 1/2	(29) 1/2	(30) 1/10
(31) 2/3	(32) 1/9	(33) 1/5
(34) 1/2	(35) 1/7	(36) 3/4
(37) 4/5	(38) 1/3	(39) 4/5
(40) 3/5	(41) 2/3	(42) 4/5
(43) 1/2	(44) 1/2	(45) 2/3

DAY 4

(1) 3/4	(2) 1/8	(3) 3/5
(4) 1/2	(5) 7/10	(6) 1/3
(7) 3/4	(8) 3/4	(9) 1/3
(10) 6/7	(11) 1/2	(12) 2/3
(13) 3/5	(14) 7/10	(15) 1/2
(16) 2/5	(17) 1/5	(18) 1/5
(19) 3/4	(20) 1/2	(21) 2/3
(22) 2/3	(23) 1/2	(24) 1/4
(25) 1/2	(26) 1/3	(27) 2/3
(28) 4/5	(29) 1/2	(30) 3/4
(31) 1/2	(32) 1/2	(33) 6/7
(34) 5/8	(35) 3/4	(36) 3/5
(37) 1/4	(38) 3/5	(39) 1/3
(40) 1/3	(41) 3/4	(42) 5/7
(43) 3/4	(44) 2/3	(45) 2/3

DAY 5

(1) 7/10	(2) 2/3	(3) 1/2
(4) 4/5	(5) 1/2	(6) 6/7
(7) 4/5	(8) 1/2	(9) 1/3
(10) 2/3	(11) 1/2	(12) 4/5
(13) 3/4	(14) 4/5	(15) 1/4
(16) 3/5	(17) 2/5	(18) 3/4
(19) 2/3	(20) 1/2	(21) 2/3
(22) 2/3	(23) 3/5	(24) 7/10
(25) 3/4	(26) 2/3	(27) 3/5
(28) 9/10	(29) 1/11	(30) 2/3
(31) 2/3	(32) 1/3	(33) 1/2
(34) 1/2	(35) 2/3	(36) 3/5
(37) 1/8	(38) 1/2	(39) 4/5
(40) 1/5	(41) 1/3	(42) 4/7
(43) 3/5	(44) 2/5	(45) 1/4

DAY 6

(1) 9/10	(2) 1/2	(3) 1/5
(4) 3/10	(5) 1/2	(6) 1/2
(7) 1/2	(8) 1/5	(9) 1/9
(10) 3/5	(11) 1/4	(12) 4/9
(13) 1/4	(14) 4/5	(15) 1/2
(16) 7/10	(17) 1/2	(18) 1/2
(19) 1/2	(20) 1/2	(21) 1/2
(22) 3/4	(23) 1/2	(24) 1/2
(25) 1/3	(26) 1/2	(27) 1/4
(28) 1/3	(29) 1/2	(30) 1/10
(31) 3/4	(32) 1/2	(33) 1/2
(34) 3/4	(35) 1/3	(36) 3/4
(37) 1/2	(38) 3/5	(39) 1/3
(40) 1/5	(41) 2/3	(42) 3/4
(43) 2/5	(44) 5/6	(45) 3/4

DAY 7

(1) 4/5	(2) 4/5	(3) 3/4
(4) 1/2	(5) 1/5	(6) 4/5
(7) 4/5	(8) 1/2	(9) 4/5
(10) 5/7	(11) 1/9	(12) 4/5
(13) 3/4	(14) 1/5	(15) 6/7
(16) 4/5	(17) 1/10	(18) 8/9
(19) 3/4	(20) 1/2	(21) 2/5
(22) 1/5	(23) 1/4	(24) 1/2
(25) 2/5	(26) 7/10	(27) 7/9
(28) 1/9	(29) 3/5	(30) 1/3
(31) 1/2	(32) 3/4	(33) 7/8
(34) 3/7	(35) 7/10	(36) 1/4
(37) 1/3	(38) 1/7	(39) 1/5
(40) 3/5	(41) 1/3	(42) 2/5
(43) 7/8	(44) 1/3	(45) 1/2

DAY 8

(1) 4/5	(2) 1/5	(3) 8/9
(4) 1/8	(5) 2/3	(6) 1/2
(7) 1/3	(8) 1/5	(9) 3/5
(10) 1/4	(11) 2/3	(12) 1/2
(13) 1/4	(14) 1/3	(15) 1/2
(16) 2/5	(17) 1/3	(18) 1/5
(19) 2/3	(20) 3/4	(21) 9/10
(22) 1/7	(23) 3/4	(24) 2/3
(25) 1/4	(26) 3/4	(27) 1/2
(28) 1/2	(29) 1/4	(30) 1/4
(31) 1/4	(32) 4/5	(33) 7/8
(34) 3/5	(35) 4/5	(36) 3/4
(37) 1/3	(38) 3/4	(39) 2/3
(40) 3/5	(41) 7/8	(42) 1/4
(43) 1/5	(44) 8/9	(45) 1/3

DAY 9

(1) 1/3	(2) 3/4	(3) 4/7
(4) 1/2	(5) 1/6	(6) 3/4
(7) 1/9	(8) 1/4	(9) 4/5
(10) 1/4	(11) 1/10	(12) 1/7
(13) 2/3	(14) 1/8	(15) 1/2
(16) 3/4	(17) 7/8	(18) 1/5
(19) 1/2	(20) 2/3	(21) 3/4
(22) 3/4	(23) 1/2	(24) 6/11
(25) 1/4	(26) 1/7	(27) 1/3
(28) 3/4	(29) 2/7	(30) 5/8
(31) 3/5	(32) 1/6	(33) 1/2
(34) 2/3	(35) 1/3	(36) 4/5
(37) 7/9	(38) 1/4	(39) 4/5
(40) 1/4	(41) 1/5	(42) 9/10
(43) 3/5	(44) 1/2	(45) 5/8

DAY 10

(1) 1/2	(2) 9/10	(3) 7/8
(4) 3/4	(5) 9/10	(6) 1/2
(7) 9/10	(8) 3/5	(9) 5/9
(10) 2/3	(11) 4/7	(12) 2/3
(13) 1/8	(14) 3/5	(15) 5/6
(16) 1/5	(17) 1/2	(18) 1/3
(19) 1/2	(20) 1/3	(21) 1/3
(22) 1/2	(23) 1/4	(24) 2/5
(25) 3/5	(26) 7/10	(27) 3/7
(28) 1/2	(29) 3/5	(30) 4/5
(31) 1/4	(32) 2/5	(33) 1/2
(34) 1/3	(35) 4/7	(36) 1/3
(37) 1/2	(38) 1/3	(39) 1/2
(40) 4/5	(41) 3/4	(42) 1/5
(43) 3/4	(44) 7/9	(45) 1/4

Equivalent Fractions – Answer Key (10 Days)

DAY 1

(1) 16/44	(2) 3/33	(3) 3/30
(4) 3/27	(5) 50/60	(6) 8/12
(7) 1/6	(8) 16/36	(9) 12/32
(10) 8/28	(11) 14/28	(12) 4/12
(13) 4/22	(14) 16/28	(15) 16/28
(16) 1/8	(17) 5/25	(18) 9/15
(19) 16/24	(20) 4/20	(21) 49/70
(22) 49/77	(23) 4/16	(24) 25/45
(25) 18/60	(26) 25/45	(27) 4/16
(28) 20/24	(29) 8/16	(30) 9/30
(31) 14/24	(32) 6/21	(33) 49/84
(34) 5/25	(35) 18/48	(36) 36/42
(37) 49/84	(38) 9/18	(39) 2/8
(40) 6/9	(41) 4/14	(42) 49/56
(43) 8/20	(44) 2/6	(45) 6/15

DAY 4

(1) 2/16	(2) 25/40	(3) 20/28
(4) 49/56	(5) 20/32	(6) 3/9
(7) 4/12	(8) 6/16	(9) 2/20
(10) 5/10	(11) 9/36	(12) 5/25
(13) 9/30	(14) 12/18	(15) 1/12
(16) 16/40	(17) 9/15	(18) 9/18
(19) 9/24	(20) 6/18	(21) 4/12
(22) 3/6	(23) 1/9	(24) 16/36
(25) 2/20	(26) 20/28	(27) 12/18
(28) 12/18	(29) 3/9	(30) 8/36
(31) 36/84	(32) 20/50	(33) 1/9
(34) 35/42	(35) 6/24	(36) 4/12
(37) 20/45	(38) 3/30	(39) 4/8
(40) 5/15	(41) 8/24	(42) 16/44
(43) 18/30	(44) 3/33	(45) 12/24

DAY 7

(1) 16/32	(2) 4/20	(3) 15/20
(4) 20/25	(5) 25/40	(6) 15/50
(7) 50/60	(8) 6/9	(9) 25/60
(10) 24/64	(11) 4/22	(12) 49/56
(13) 16/20	(14) 1/12	(15) 32/72
(16) 8/28	(17) 9/45	(18) 49/84
(19) 8/12	(20) 16/36	(21) 9/33
(22) 10/24	(23) 25/60	(24) 28/36
(25) 9/18	(26) 3/15	(27) 16/44
(28) 3/18	(29) 2/14	(30) 1/6
(31) 4/18	(32) 2/14	(33) 25/100
(34) 16/44	(35) 1/4	(36) 3/9
(37) 49/70	(38) 25/30	(39) 18/48
(40) 16/28	(41) 8/12	(42) 18/24
(43) 16/56	(44) 12/27	(45) 4/18

DAY 10

(1) 8/20	(2) 40/64	(3) 24/64
(4) 4/18	(5) 2/14	(6) 9/33
(7) 9/27	(8) 24/64	(9) 40/64
(10) 3/21	(11) 25/45	(12) 30/36
(13) 49/70	(14) 49/70	(15) 42/49
(16) 2/8	(17) 35/49	(18) 20/32
(19) 8/10	(20) 3/21	(21) 24/32
(22) 6/24	(23) 25/50	(24) 9/33
(25) 49/84	(26) 6/12	(27) 50/80
(28) 9/27	(29) 4/16	(30) 3/12
(31) 3/18	(32) 6/12	(33) 8/32
(34) 12/24	(35) 10/16	(36) 25/40
(37) 3/27	(38) 3/21	(39) 25/30
(40) 15/30	(41) 5/10	(42) 20/100
(43) 25/55	(44) 9/12	(45) 35/49

DAY 2

(1) 14/24	(2) 9/33	(3) 12/20
(4) 1/9	(5) 12/48	(6) 36/84
(7) 1/12	(8) 3/33	(9) 4/14
(10) 25/60	(11) 36/45	(12) 16/28
(13) 8/48	(14) 8/36	(15) 25/30
(16) 4/8	(17) 49/56	(18) 2/12
(19) 42/49	(20) 16/24	(21) 3/12
(22) 5/15	(23) 20/35	(24) 12/32
(25) 16/28	(26) 72/90	(27) 12/18
(28) 24/64	(29) 36/84	(30) 16/36
(31) 4/8	(32) 25/30	(33) 49/70
(34) 2/16	(35) 9/27	(36) 9/33
(37) 50/80	(38) 18/24	(39) 28/36
(40) 7/35	(41) 25/63	(42) 12/30
(43) 9/36	(44) 49/84	(45) 9/30

DAY 5

(1) 12/16	(2) 7/28	(3) 25/50
(4) 4/10	(5) 16/20	(6) 25/55
(7) 25/30	(8) 1/12	(9) 20/50
(10) 4/12	(11) 5/20	(12) 8/16
(13) 16/44	(14) 16/28	(15) 9/15
(16) 2/12	(17) 3/12	(18) 9/27
(19) 12/72	(20) 63/99	(21) 2/16
(22) 40/64	(23) 49/56	(24) 12/30
(25) 9/18	(26) 32/56	(27) 5/25
(28) 4/18	(29) 17/34	(30) 4/12
(31) 12/18	(32) 16/24	(33) 9/33
(34) 1/10	(35) 1/12	(36) 49/63
(37) 25/40	(38) 4/24	(39) 3/27
(40) 12/20	(41) 49/63	(42) 12/36
(43) 49/56	(44) 25/30	(45) 49/84

DAY 8

(1) 6/8	(2) 4/22	(3) 8/32
(4) 12/16	(5) 25/55	(6) 63/99
(7) 20/50	(8) 12/18	(9) 25/30
(10) 10/12	(11) 49/70	(12) 24/32
(13) 50/60	(14) 4/32	(15) 15/50
(16) 20/50	(17) 25/45	(18) 12/24
(19) 9/24	(20) 49/77	(21) 9/15
(22) 12/54	(23) 1/9	(24) 40/80
(25) 24/30	(26) 15/25	(27) 3/18
(28) 9/30	(29) 16/44	(30) 20/100
(31) 9/15	(32) 9/21	(33) 25/40
(34) 9/22	(35) 9/27	(36) 4/8
(37) 9/27	(38) 9/21	(39) 9/24
(40) 12/30	(41) 20/25	(42) 9/36
(43) 49/63	(44) 6/18	(45) 8/28

DAY 3

(1) 4/6	(2) 2/16	(3) 1/12
(4) 2/20	(5) 12/18	(6) 25/30
(7) 3/18	(8) 4/18	(9) 30/90
(10) 49/84	(11) 10/20	(12) 2/20
(13) 12/24	(14) 4/22	(15) 12/24
(16) 25/60	(17) 9/27	(18) 3/12
(19) 25/45	(20) 49/70	(21) 8/12
(22) 36/66	(23) 36/42	(24) 36/42
(25) 9/18	(26) 12/18	(27) 9/15
(28) 49/63	(29) 3/27	(30) 16/40
(31) 4/18	(32) 20/35	(33) 49/77
(34) 25/35	(35) 4/16	(36) 49/56
(37) 9/27	(38) 49/63	(39) 5/25
(40) 8/40	(41) 9/36	(42) 20/35
(43) 2/8	(44) 1/11	(45) 4/18

DAY 6

(1) 3/18	(2) 3/21	(3) 9/18
(4) 3/18	(5) 14/20	(6) 1/11
(7) 3/9	(8) 49/84	(9) 4/12
(10) 6/24	(11) 2/16	(12) 16/64
(13) 3/9	(14) 25/60	(15) 12/24
(16) 8/16	(17) 8/36	(18) 12/27
(19) 12/16	(20) 6/9	(21) 15/21
(22) 20/45	(23) 9/18	(24) 1/5
(25) 16/32	(26) 16/28	(27) 36/66
(28) 36/42	(29) 4/16	(30) 1/11
(31) 16/44	(32) 25/35	(33) 6/18
(34) 4/12	(35) 8/10	(36) 15/25
(37) 4/48	(38) 9/21	(39) 9/21
(40) 4/22	(41) 1/10	(42) 4/24
(43) 1/12	(44) 1/7	(45) 16/24

DAY 9

(1) 9/33	(2) 5/25	(3) 9/33
(4) 16/28	(5) 40/64	(6) 16/36
(7) 9/30	(8) 2/16	(9) 8/28
(10) 6/24	(11) 18/30	(12) 4/24
(13) 3/24	(14) 21/33	(15) 36/42
(16) 18/48	(17) 9/33	(18) 4/14
(19) 16/28	(20) 9/36	(21) 10/40
(22) 8/20	(23) 4/14	(24) 9/27
(25) 9/60	(26) 1/11	(27) 25/50
(28) 16/28	(29) 14/21	(30) 16/24
(31) 8/12	(32) 1/12	(33) 24/32
(34) 15/25	(35) 30/36	(36) 5/25
(37) 63/99	(38) 12/18	(39) 4/18
(40) 9/27	(41) 25/50	(42) 25/45
(43) 12/24	(44) 32/40	(45) 3/27

Decimals to Fractions – Answer Key (10 Days)

DAY 1
(1) $\frac{2}{5}$ (2) $1\frac{1}{8}$ (3) $1\frac{9}{10}$ (4) $\frac{1}{8}$ (5) $9\frac{1}{2}$ (6) $6\frac{1}{5}$ (7) $\frac{7}{8}$ (8) $3\frac{5}{8}$ (9) $3\frac{2}{5}$ (10) $\frac{3}{10}$ (11) $\frac{1}{6}$ (12) $8\frac{1}{2}$ (13) $\frac{3}{5}$ (14) $5\frac{4}{5}$ (15) $5\frac{3}{10}$ (16) $\frac{1}{5}$ (17) $2\frac{1}{4}$ (18) $2\frac{1}{10}$ (19) $\frac{2}{3}$ (20) $4\frac{2}{3}$ (21) $7\frac{7}{10}$ (22) $\frac{9}{10}$ (23) $1\frac{4}{5}$ (24) $4\frac{1}{10}$ (25) $\frac{1}{4}$ (26) $8\frac{1}{8}$ (27) $9\frac{3}{5}$ (28) $\frac{5}{8}$ (29) $3\frac{1}{5}$ (30) $6\frac{9}{10}$ (31) $\frac{1}{10}$ (32) $\frac{2}{5}$ (33) $1\frac{7}{10}$ (34) $\frac{1}{8}$ (35) $6\frac{3}{8}$ (36) $3\frac{4}{5}$ (37) $\frac{3}{8}$ (38) $2\frac{1}{5}$ (39) $8\frac{1}{5}$ (40) $\frac{7}{8}$ (41) $5\frac{1}{2}$ (42) $5\frac{9}{10}$ (43) $\frac{5}{8}$ (44) $1\frac{1}{3}$ (45) $2\frac{3}{5}$

DAY 2
(1) $\frac{1}{2}$ (2) $7\frac{1}{4}$ (3) $7\frac{3}{10}$ (4) $\frac{1}{3}$ (5) $4\frac{4}{5}$ (6) $4\frac{4}{5}$ (7) $\frac{3}{5}$ (8) $\frac{1}{8}$ (9) $9\frac{4}{5}$ (10) $\frac{1}{8}$ (11) $3\frac{1}{3}$ (12) $5\frac{1}{2}$ (13) $\frac{1}{5}$ (14) $2\frac{4}{5}$ (15) $1\frac{3}{10}$ (16) $\frac{5}{8}$ (17) $6\frac{1}{8}$ (18) $3\frac{1}{5}$ (19) $\frac{3}{8}$ (20) $1\frac{5}{8}$ (21) $8\frac{4}{5}$ (22) $\frac{9}{10}$ (23) $8\frac{3}{4}$ (24) $6\frac{1}{2}$ (25) $\frac{1}{2}$ (26) $4\frac{1}{8}$ (27) $2\frac{9}{10}$ (28) $\frac{5}{8}$ (29) $5\frac{2}{3}$ (30) $7\frac{1}{5}$ (31) $\frac{1}{4}$ (32) $2\frac{3}{8}$ (33) $4\frac{9}{10}$ (34) $\frac{3}{5}$ (35) $7\frac{1}{5}$ (36) $9\frac{7}{10}$ (37) $\frac{3}{4}$ (38) $2\frac{2}{5}$ (39) $5\frac{1}{10}$ (40) $\frac{1}{8}$ (41) $\frac{3}{10}$ (42) $1\frac{2}{5}$ (43) $\frac{3}{8}$ (44) $3\frac{3}{5}$ (45) $3\frac{3}{5}$

DAY 3
(1) $\frac{7}{8}$ (2) $1\frac{3}{5}$ (3) $8\frac{9}{10}$ (4) $\frac{5}{8}$ (5) $2\frac{1}{5}$ (6) $6\frac{2}{5}$ (7) $\frac{1}{8}$ (8) $\frac{9}{10}$ (9) $2\frac{3}{10}$ (10) $\frac{1}{2}$ (11) $1\frac{3}{10}$ (12) $7\frac{4}{5}$ (13) $\frac{5}{8}$ (14) $3\frac{1}{3}$ (15) $4\frac{1}{2}$ (16) $\frac{1}{4}$ (17) $1\frac{7}{10}$ (18) $9\frac{1}{2}$ (19) $\frac{3}{4}$ (20) $2\frac{1}{2}$ (21) $5\frac{7}{10}$ (22) $\frac{1}{8}$ (23) $\frac{3}{5}$ (24) $1\frac{1}{5}$ (25) $\frac{3}{8}$ (26) $3\frac{4}{5}$ (27) $3\frac{7}{10}$ (28) $\frac{7}{8}$ (29) $1\frac{1}{2}$ (30) $8\frac{7}{10}$ (31) $2\frac{1}{2}$ (32) $2\frac{1}{10}$ (33) $6\frac{3}{10}$ (34) $1\frac{7}{10}$ (35) $\frac{2}{5}$ (36) $2\frac{4}{5}$ (37) $2\frac{1}{2}$ (38) $3\frac{1}{5}$ (39) $7\frac{3}{5}$ (40) $3\frac{9}{10}$ (41) $1\frac{1}{2}$ (42) $4\frac{1}{5}$ (43) $6\frac{1}{4}$ (44) $2\frac{9}{10}$ (45) $9\frac{3}{10}$

DAY 4
(1) $2\frac{3}{4}$ (2) $\frac{4}{5}$ (3) $2\frac{3}{4}$ (4) $1\frac{1}{2}$ (5) $3\frac{2}{5}$ (6) $3\frac{3}{5}$ (7) $3\frac{1}{5}$ (8) $1\frac{2}{5}$ (9) $4\frac{1}{4}$ (10) $7\frac{1}{8}$ (11) $2\frac{3}{10}$ (12) $5\frac{1}{8}$ (13) $\frac{3}{4}$ (14) $\frac{1}{5}$ (15) $6\frac{2}{5}$ (16) $9\frac{5}{8}$ (17) $3\frac{1}{10}$ (18) $7\frac{3}{4}$ (19) $4\frac{1}{3}$ (20) $1\frac{1}{10}$ (21) $8\frac{1}{3}$ (22) $5\frac{1}{2}$ (23) $2\frac{3}{5}$ (24) $9\frac{7}{8}$ (25) $2\frac{7}{8}$ (26) $\frac{7}{10}$ (27) $10\frac{1}{5}$ (28) $\frac{3}{8}$ (29) $3\frac{7}{10}$ (30) $1\frac{3}{10}$ (31) $6\frac{4}{5}$ (32) $1\frac{4}{5}$ (33) $2\frac{1}{2}$ (34) $\frac{3}{4}$ (35) $2\frac{4}{5}$ (36) $3\frac{1}{5}$ (37) $2\frac{3}{5}$ (38) $\frac{1}{2}$ (39) $4\frac{2}{5}$ (40) $4\frac{1}{8}$ (41) $3\frac{1}{5}$ (42) $5\frac{7}{10}$ (43) $\frac{4}{5}$ (44) $1\frac{9}{10}$ (45) $6\frac{1}{10}$

DAY 5
(1) $8\frac{5}{8}$ (2) $2\frac{7}{10}$ (3) $7\frac{4}{5}$ (4) $3\frac{1}{5}$ (5) $\frac{1}{10}$ (6) $8\frac{9}{10}$ (7) $5\frac{1}{4}$ (8) $3\frac{9}{10}$ (9) $1\frac{3}{5}$ (10) $2\frac{1}{8}$ (11) $4\frac{3}{5}$ (12) $2\frac{9}{10}$ (13) $\frac{5}{8}$ (14) $8\frac{2}{5}$ (15) $3\frac{1}{2}$ (16) $7\frac{1}{2}$ (17) $5\frac{1}{2}$ (18) $4\frac{1}{10}$ (19) $1\frac{3}{5}$ (20) $9\frac{1}{2}$ (21) $5\frac{3}{10}$ (22) $3\frac{3}{5}$ (23) $6\frac{4}{5}$ (24) $6\frac{7}{10}$ (25) $\frac{3}{5}$ (26) $2\frac{9}{10}$ (27) $7\frac{1}{5}$ (28) $9\frac{3}{4}$ (29) $7\frac{1}{10}$ (30) $8\frac{1}{2}$ (31) $2\frac{1}{3}$ (32) $1\frac{3}{5}$ (33) $1\frac{9}{10}$ (34) $6\frac{3}{5}$ (35) $3\frac{1}{3}$ (36) $2\frac{7}{10}$ (37) $1\frac{4}{5}$ (38) $8\frac{3}{5}$ (39) $3\frac{3}{5}$ (40) $4\frac{1}{5}$ (41) $6\frac{3}{10}$ (42) $4\frac{3}{10}$ (43) $\frac{7}{8}$ (44) $2\frac{1}{5}$ (45) $5\frac{3}{5}$

DAY 6
(1) $5\frac{1}{2}$ (2) $7\frac{4}{5}$ (3) $6\frac{2}{5}$ (4) $7\frac{3}{8}$ (5) $4\frac{1}{5}$ (6) $7\frac{1}{10}$ (7) $1\frac{1}{4}$ (8) $9\frac{1}{10}$ (9) $8\frac{4}{5}$ (10) $2\frac{7}{8}$ (11) $5\frac{7}{10}$ (12) $1\frac{2}{5}$ (13) $2\frac{1}{8}$ (14) $1\frac{1}{5}$ (15) $2\frac{1}{5}$ (16) $3\frac{2}{5}$ (17) $3\frac{7}{10}$ (18) $3\frac{9}{10}$ (19) $\frac{1}{4}$ (20) $8\frac{3}{10}$ (21) $4\frac{1}{2}$ (22) $4\frac{4}{5}$ (23) $6\frac{1}{10}$ (24) $5\frac{1}{5}$ (25) $1\frac{5}{8}$ (26) $2\frac{4}{5}$ (27) $6\frac{3}{10}$ (28) $2\frac{1}{4}$ (29) $7\frac{1}{10}$ (30) $7\frac{3}{5}$ (31) $7\frac{1}{5}$ (32) $4\frac{9}{10}$ (33) $8\frac{1}{5}$ (34) $5\frac{1}{2}$ (35) $9\frac{7}{10}$ (36) $1\frac{7}{10}$ (37) $\frac{1}{2}$ (38) $5\frac{1}{10}$ (39) $2\frac{3}{10}$ (40) $3\frac{3}{5}$ (41) $1\frac{2}{5}$ (42) $3\frac{4}{5}$ (43) $6\frac{1}{8}$ (44) $3\frac{3}{5}$ (45) $4\frac{9}{10}$

DAY 7
(1) $\frac{7}{8}$ (2) $8\frac{9}{10}$ (3) $5\frac{2}{5}$ (4) $1\frac{2}{5}$ (5) $6\frac{2}{5}$ (6) $6\frac{9}{10}$ (7) $8\frac{1}{4}$ (8) $2\frac{3}{10}$ (9) $7\frac{3}{10}$ (10) $2\frac{4}{5}$ (11) $7\frac{4}{5}$ (12) $8\frac{3}{5}$ (13) $5\frac{7}{8}$ (14) $4\frac{2}{5}$ (15) $1\frac{1}{5}$ (16) $1\frac{1}{8}$ (17) $9\frac{1}{5}$ (18) $2\frac{2}{5}$ (19) $7\frac{1}{2}$ (20) $5\frac{4}{5}$ (21) $3\frac{2}{5}$ (22) $2\frac{1}{6}$ (23) $1\frac{1}{10}$ (24) $4\frac{4}{5}$ (25) $3\frac{3}{4}$ (26) $3\frac{1}{2}$ (27) $5\frac{1}{5}$ (28) $1\frac{1}{8}$ (29) $8\frac{7}{10}$ (30) $6\frac{3}{5}$ (31) $4\frac{5}{8}$ (32) $6\frac{1}{2}$ (33) $7\frac{4}{5}$ (34) $\frac{1}{3}$ (35) $2\frac{1}{2}$ (36) $8\frac{1}{3}$ (37) $4\frac{1}{2}$ (38) $7\frac{1}{2}$ (39) $1\frac{4}{5}$ (40) $9\frac{4}{5}$ (41) $4\frac{1}{2}$ (42) $2\frac{4}{5}$ (43) $2\frac{1}{8}$ (44) $2\frac{2}{5}$ (45) $1\frac{7}{10}$

DAY 8
(1) $\frac{5}{8}$ (2) $7\frac{2}{5}$ (3) $2\frac{3}{10}$ (4) $5\frac{2}{5}$ (5) $4\frac{7}{10}$ (6) $3\frac{9}{10}$ (7) $1\frac{7}{8}$ (8) $9\frac{2}{5}$ (9) $4\frac{1}{5}$ (10) $3\frac{1}{8}$ (11) $5\frac{3}{5}$ (12) $5\frac{2}{5}$ (13) $6\frac{3}{8}$ (14) $1\frac{9}{10}$ (15) $6\frac{9}{10}$ (16) $\frac{7}{8}$ (17) $6\frac{1}{8}$ (18) $7\frac{3}{10}$ (19) $2\frac{1}{5}$ (20) $3\frac{2}{5}$ (21) $8\frac{3}{5}$ (22) $4\frac{1}{4}$ (23) $8\frac{1}{2}$ (24) $1\frac{1}{2}$ (25) $1\frac{1}{3}$ (26) $5\frac{3}{10}$ (27) $2\frac{4}{5}$ (28) $8\frac{1}{2}$ (29) $2\frac{1}{10}$ (30) $3\frac{2}{5}$ (31) $3\frac{5}{8}$ (32) $7\frac{7}{10}$ (33) $4\frac{4}{5}$ (34) $\frac{1}{5}$ (35) $4\frac{1}{10}$ (36) $5\frac{1}{2}$ (37) $6\frac{4}{5}$ (38) $9\frac{3}{5}$ (39) $6\frac{3}{5}$ (40) $2\frac{5}{8}$ (41) $6\frac{9}{10}$ (42) $7\frac{4}{5}$ (43) $5\frac{1}{5}$ (44) $1\frac{7}{10}$ (45) $8\frac{1}{3}$

DAY 9
(1) $1\frac{1}{2}$ (2) $3\frac{4}{5}$ (3) $1\frac{9}{10}$ (4) $4\frac{2}{3}$ (5) $8\frac{1}{6}$ (6) $2\frac{7}{10}$ (7) $\frac{3}{4}$ (8) $5\frac{9}{10}$ (9) $3\frac{3}{5}$ (10) $1\frac{1}{4}$ (11) $2\frac{3}{5}$ (12) $4\frac{3}{10}$ (13) $3\frac{4}{5}$ (14) $7\frac{3}{10}$ (15) $5\frac{3}{5}$ (16) $2\frac{3}{8}$ (17) $4\frac{4}{5}$ (18) $6\frac{2}{5}$ (19) $5\frac{5}{8}$ (20) $9\frac{4}{5}$ (21) $7\frac{1}{10}$ (22) $1\frac{3}{5}$ (23) $5\frac{1}{5}$ (24) $8\frac{4}{5}$ (25) $2\frac{3}{8}$ (26) $1\frac{3}{10}$ (27) $1\frac{2}{5}$ (28) $\frac{4}{5}$ (29) $3\frac{1}{5}$ (30) $2\frac{1}{5}$ (31) $3\frac{1}{8}$ (32) $8\frac{4}{5}$ (33) $3\frac{4}{5}$ (34) $5\frac{2}{5}$ (35) $6\frac{1}{2}$ (36) $4\frac{9}{10}$ (37) $1\frac{1}{5}$ (38) $2\frac{9}{10}$ (39) $5\frac{1}{5}$ (40) $7\frac{5}{8}$ (41) $7\frac{1}{5}$ (42) $6\frac{3}{10}$ (43) $4\frac{1}{2}$ (44) $4\frac{9}{10}$ (45) $7\frac{3}{5}$

DAY 10
(1) $\frac{7}{8}$ (2) $9\frac{7}{10}$ (3) $8\frac{1}{5}$ (4) $2\frac{1}{5}$ (5) $5\frac{1}{10}$ (6) $1\frac{3}{5}$ (7) $6\frac{3}{4}$ (8) $1\frac{2}{5}$ (9) $2\frac{1}{5}$ (10) $1\frac{5}{8}$ (11) $3\frac{3}{5}$ (12) $3\frac{1}{2}$ (13) $8\frac{1}{5}$ (14) $8\frac{9}{10}$ (15) $4\frac{1}{10}$ (16) $3\frac{4}{5}$ (17) $6\frac{1}{5}$ (18) $5\frac{3}{10}$ (19) $\frac{1}{4}$ (20) $2\frac{3}{10}$ (21) $6\frac{7}{10}$ (22) $5\frac{1}{8}$ (23) $7\frac{4}{5}$ (24) $7\frac{1}{5}$ (25) $2\frac{3}{5}$ (26) $4\frac{1}{2}$ (27) $8\frac{1}{2}$ (28) $4\frac{1}{4}$ (29) $9\frac{1}{2}$ (30) $1\frac{3}{10}$ (31) $1\frac{2}{5}$ (32) $2\frac{2}{5}$ (33) $1\frac{1}{2}$ (34) $7\frac{7}{10}$ (35) $7\frac{2}{5}$ (36) $3\frac{1}{5}$ (37) $\frac{3}{5}$ (38) $4\frac{7}{10}$ (39) $4\frac{2}{5}$ (40) $2\frac{7}{8}$ (41) $9\frac{2}{5}$ (42) $5\frac{7}{10}$ (43) $6\frac{2}{5}$ (44) $5\frac{3}{5}$ (45) $6\frac{1}{10}$

Adding Fractions - Answer Key

DAY 1
(1) 5/5 (2) 10/7 (3) 4/7
(4) 3/6 (5) 4/8 (6) 4/9
(7) 9/9 (8) 6/10 (9) 4/12
(10) 11/12 (11) 8/12 (12) 8/8
(13) 3/3 (14) 2/5 (15) 4/8
(16) 2/7 (17) 10/9 (18) 3/8
(19) 11/10 (20) 4/9 (21) 8/8
(22) 7/8 (23) 6/12 (24) 7/10
(25) 7/7 (26) 8/6 (27) 4/3
(28) 4/4 (29) 2/12 (30) 5/8
(31) 5/5 (32) 10/10 (33) 6/5
(34) 6/7 (35) 4/12 (36) 7/7
(37) 8/8 (38) 6/7 (39) 2/6
(40) 2/3 (41) 8/8 (42) 7/7
(43) 9/6 (44) 2/7 (45) 8/12

DAY 4
(1) 5/6 (2) 5/5 (3) 6/12
(4) 5/6 (5) 6/9 (6) 5/3
(7) 11/12 (8) 7/7 (9) 9/7
(10) 5/7 (11) 5/12 (12) 10/7
(13) 6/9 (14) 5/4 (15) 4/9
(16) 5/10 (17) 5/10 (18) 6/9
(19) 7/10 (20) 6/6 (21) 6/8
(22) 8/7 (23) 7/10 (24) 10/10
(25) 3/3 (26) 5/5 (27) 11/12
(28) 3/4 (29) 6/8 (30) 8/8
(31) 5/8 (32) 5/7 (33) 6/6
(34) 5/7 (35) 6/12 (36) 4/3
(37) 18/10 (38) 8/9 (39) 10/10
(40) 3/5 (41) 6/8 (42) 7/7
(43) 9/9 (44) 7/3 (45) 7/12

DAY 7
(1) 9/9 (2) 6/10 (3) 11/10
(4) 4/10 (5) 6/6 (6) 6/10
(7) 10/10 (8) 7/9 (9) 11/10
(10) 6/8 (11) 7/8 (12) 5/12
(13) 2/3 (14) 4/5 (15) 5/10
(16) 4/3 (17) 4/8 (18) 11/12
(19) 8/12 (20) 6/10 (21) 5/7
(22) 5/9 (23) 5/5 (24) 8/8
(25) 3/10 (26) 8/9 (27) 8/10
(28) 7/7 (29) 7/12 (30) 5/5
(31) 5/9 (32) 9/8 (33) 9/9
(34) 9/10 (35) 5/6 (36) 5/3
(37) 3/4 (38) 10/10 (39) 4/5
(40) 6/5 (41) 7/7 (42) 10/10
(43) 9/8 (44) 6/6 (45) 8/5

DAY 10
(1) 6/6 (2) 6/12 (3) 7/6
(4) 2/7 (5) 6/7 (6) 8/5
(7) 4/5 (8) 8/6 (9) 9/10
(10) 6/9 (11) 9/9 (12) 9/7
(13) 5/5 (14) 4/10 (15) 6/7
(16) 7/7 (17) 9/5 (18) 7/12
(19) 6/10 (20) 6/8 (21) 8/9
(22) 6/12 (23) 4/5 (24) 11/8
(25) 4/10 (26) 8/6 (27) 6/10
(28) 6/6 (29) 6/12 (30) 8/9
(31) 4/7 (32) 3/4 (33) 7/3
(34) 4/9 (35) 7/9 (36) 3/3
(37) 4/12 (38) 8/10 (39) 3/5
(40) 8/10 (41) 7/7 (42) 6/6
(43) 4/10 (44) 7/8 (45) 4/4

DAY 2
(1) 5/4 (2) 10/6 (3) 7/4
(4) 9/8 (5) 4/10 (6) 2/8
(7) 6/9 (8) 6/12 (9) 4/8
(10) 8/10 (11) 8/9 (12) 6/9
(13) 4/5 (14) 2/4 (15) 3/5
(16) 6/6 (17) 10/8 (18) 8/10
(19) 7/10 (20) 4/5 (21) 5/3
(22) 2/2 (23) 6/5 (24) 5/7
(25) 11/12 (26) 8/7 (27) 3/9
(28) 5/7 (29) 2/10 (30) 5/6
(31) 4/6 (32) 10/12 (33) 4/12
(34) 8/10 (35) 4/4 (36) 8/12
(37) 3/3 (38) 6/9 (39) 3/6
(40) 11/9 (41) 8/12 (42) 4/4
(43) 7/8 (44) 2/4 (45) 6/9

DAY 5
(1) 5/6 (2) 8/5 (3) 10/8
(4) 4/9 (5) 5/10 (6) 7/4
(7) 7/12 (8) 6/3 (9) 5/7
(10) 2/2 (11) 7/9 (12) 6/6
(13) 10/8 (14) 4/6 (15) 9/10
(16) 5/6 (17) 6/12 (18) 8/10
(19) 9/7 (20) 4/12 (21) 10/12
(22) 6/5 (23) 5/7 (24) 4/6
(25) 4/4 (26) 6/4 (27) 10/4
(28) 9/9 (29) 6/7 (30) 5/5
(31) 6/7 (32) 3/9 (33) 12/9
(34) 3/3 (35) 5/6 (36) 12/12
(37) 9/10 (38) 6/12 (39) 5/8
(40) 4/5 (41) 6/8 (42) 10/9
(43) 12/12 (44) 5/7 (45) 8/5

DAY 8
(1) 5/10 (2) 7/12 (3) 3/3
(4) 4/7 (5) 5/5 (6) 5/7
(7) 10/10 (8) 8/4 (9) 8/12
(10) 5/5 (11) 7/5 (12) 10/8
(13) 10/12 (14) 5/7 (15) 4/4
(16) 9/9 (17) 7/7 (18) 6/4
(19) 10/10 (20) 5/10 (21) 6/6
(22) 5/4 (23) 5/12 (24) 4/2
(25) 4/7 (26) 8/4 (27) 7/5
(28) 4/2 (29) 7/7 (30) 6/6
(31) 7/10 (32) 7/7 (33) 7/12
(34) 2/8 (35) 6/9 (36) 7/7
(37) 6/12 (38) 4/4 (39) 10/9
(40) 4/3 (41) 8/5 (42) 10/8
(43) 8/8 (44) 6/6 (45) 5/10

DAY 11
(1) 3/8 (2) 6/10 (3) 7/7
(4) 8/8 (5) 3/6 (6) 5/9
(7) 7/9 (8) 9/10 (9) 4/8
(10) 4/3 (11) 6/9 (12) 5/10
(13) 5/4 (14) 5/8 (15) 7/12
(16) 6/5 (17) 6/6 (18) 6/10
(19) 8/7 (20) 3/12 (21) 5/9
(22) 2/6 (23) 6/4 (24) 8/8
(25) 7/7 (26) 4/7 (27) 4/10
(28) 8/12 (29) 6/5 (30) 5/6
(31) 7/4 (32) 4/9 (33) 4/7
(34) 2/10 (35) 3/10 (36) 10/10
(37) 4/8 (38) 10/7 (39) 5/5
(40) 6/9 (41) 8/6 (42) 10/12
(43) 3/5 (44) 5/8 (45) 7/7

DAY 3
(1) 6/10 (2) 10/9 (3) 5/3
(4) 4/4 (5) 4/6 (6) 7/7
(7) 4/8 (8) 6/6 (9) 11/8
(10) 8/10 (11) 8/10 (12) 8/8
(13) 3/7 (14) 2/6 (15) 9/8
(16) 6/6 (17) 5/6 (18) 6/5
(19) 9/12 (20) 4/6 (21) 8/5
(22) 3/9 (23) 3/8 (24) 7/12
(25) 5/5 (26) 3/10 (27) 11/6
(28) 7/7 (29) 6/9 (30) 9/10
(31) 3/8 (32) 6/7 (33) 9/5
(34) 10/10 (35) 8/8 (36) 5/8
(37) 4/12 (38) 5/12 (39) 13/9
(40) 8/9 (41) 6/12 (42) 6/4
(43) 3/10 (44) 8/12 (45) 5/10

DAY 6
(1) 5/6 (2) 5/12 (3) 11/9
(4) 9/10 (5) 8/6 (6) 4/8
(7) 2/5 (8) 7/4 (9) 5/12
(10) 6/8 (11) 8/10 (12) 8/5
(13) 6/7 (14) 5/5 (15) 6/4
(16) 4/8 (17) 6/5 (18) 8/8
(19) 10/9 (20) 6/10 (21) 5/4
(22) 6/7 (23) 5/12 (24) 11/5
(25) 6/4 (26) 4/7 (27) 11/10
(28) 9/7 (29) 6/3 (30) 9/6
(31) 4/4 (32) 7/5 (33) 7/7
(34) 3/6 (35) 5/12 (36) 7/7
(37) 7/10 (38) 7/8 (39) 7/9
(40) 8/6 (41) 7/6 (42) 8/4
(43) 5/8 (44) 5/4 (45) 7/6

DAY 9
(1) 6/6 (2) 6/10 (3) 11/10
(4) 6/10 (5) 7/8 (6) 4/7
(7) 12/9 (8) 7/12 (9) 9/10
(10) 5/5 (11) 8/8 (12) 10/10
(13) 6/6 (14) 5/12 (15) 6/5
(16) 3/3 (17) 4/7 (18) 10/9
(19) 6/7 (20) 9/9 (21) 6/3
(22) 4/4 (23) 8/10 (24) 7/8
(25) 6/8 (26) 4/6 (27) 5/5
(28) 6/4 (29) 8/12 (30) 5/4
(31) 5/9 (32) 5/7 (33) 8/12
(34) 6/10 (35) 4/8 (36) 6/7
(37) 7/10 (38) 7/6 (39) 7/12
(40) 4/6 (41) 4/12 (42) 5/3
(43) 6/8 (44) 9/4 (45) 5/4

DAY 12
(1) 8/10 (2) 3/9 (3) 11/10
(4) 5/3 (5) 9/10 (6) 7/8
(7) 7/7 (8) 7/5 (9) 7/9
(10) 3/9 (11) 3/8 (12) 7/7
(13) 5/6 (14) 9/10 (15) 5/5
(16) 4/12 (17) 6/10 (18) 7/8
(19) 8/10 (20) 3/4 (21) 11/10
(22) 5/7 (23) 8/8 (24) 4/10
(25) 6/6 (26) 5/6 (27) 7/6
(28) 4/3 (29) 3/7 (30) 9/9
(31) 7/8 (32) 9/9 (33) 5/5
(34) 6/5 (35) 8/10 (36) 7/10
(37) 5/2 (38) 6/7 (39) 7/10
(40) 9/7 (41) 3/6 (42) 5/12
(43) 5/10 (44) 9/12 (45) 4/12

Adding Fractions – Answer Key (25 Days)

DAY 13
(1) 4/7 (2) 5/10 (3) 7/10
(4) 7/6 (5) 3/10 (6) 6/8
(7) 7/5 (8) 7/7 (9) 7/12
(10) 4/8 (11) 5/3 (12) 3/4
(13) 9/12 (14) 3/12 (15) 3/10
(16) 5/6 (17) 9/8 (18) 5/6
(19) 4/8 (20) 5/4 (21) 3/3
(22) 8/10 (23) 3/5 (24) 4/4
(25) 5/8 (26) 9/10 (27) 10/8
(28) 3/6 (29) 5/9 (30) 8/9
(31) 9/12 (32) 3/2 (33) 5/5
(34) 5/10 (35) 7/6 (36) 5/6
(37) 3/12 (38) 5/12 (39) 8/10
(40) 6/9 (41) 4/10 (42) 3/12
(43) 7/4 (44) 8/6 (45) 6/6

DAY 14
(1) 3/12 (2) 4/7 (3) 8/8
(4) 9/8 (5) 3/3 (6) 9/7
(7) 5/9 (8) 8/9 (9) 10/12
(10) 3/10 (11) 6/5 (12) 5/5
(13) 9/12 (14) 4/10 (15) 9/9
(16) 7/7 (17) 8/7 (18) 10/10
(19) 7/5 (20) 6/8 (21) 6/3
(22) 4/4 (23) 4/6 (24) 6/8
(25) 7/6 (26) 8/5 (27) 3/9
(28) 6/7 (29) 7/10 (30) 7/10
(31) 2/9 (32) 4/4 (33) 10/10
(34) 7/10 (35) 8/9 (36) 2/2
(37) 5/12 (38) 6/9 (39) 6/6
(40) 3/7 (41) 4/9 (42) 3/8
(43) 9/12 (44) 8/7 (45) 4/4

DAY 15
(1) 6/6 (2) 6/4 (3) 5/10
(4) 3/3 (5) 4/12 (6) 9/9
(7) 8/8 (8) 8/8 (9) 4/7
(10) 5/4 (11) 6/10 (12) 7/12
(13) 2/5 (14) 4/10 (15) 9/8
(16) 8/9 (17) 7/8 (18) 3/5
(19) 6/10 (20) 7/5 (21) 6/6
(22) 4/12 (23) 4/7 (24) 2/3
(25) 6/7 (26) 7/10 (27) 6/7
(28) 6/8 (29) 6/9 (30) 4/4
(31) 4/6 (32) 5/3 (33) 3/7
(34) 9/12 (35) 6/9 (36) 7/5
(37) 4/7 (38) 2/10 (39) 5/12
(40) 9/10 (41) 8/7 (42) 3/6
(43) 2/12 (44) 6/6 (45) 6/10

DAY 16
(1) 8/9 (2) 3/5 (3) 2/5
(4) 5/5 (5) 8/8 (6) 9/12
(7) 3/8 (8) 7/4 (9) 5/9
(10) 4/7 (11) 4/12 (12) 3/4
(13) 6/12 (14) 7/10 (15) 5/7
(16) 6/3 (17) 6/8 (18) 7/7
(19) 5/4 (20) 2/7 (21) 2/10
(22) 4/5 (23) 9/5 (24) 5/8
(25) 6/7 (26) 6/10 (27) 6/6
(28) 2/12 (29) 2/6 (30) 2/7
(31) 2/10 (32) 5/9 (33) 3/9
(34) 10/9 (35) 5/7 (36) 7/12
(37) 4/4 (38) 2/10 (39) 9/12
(40) 6/8 (41) 9/6 (42) 5/12
(43) 6/12 (44) 6/12 (45) 3/12

DAY 17
(1) 2/7 (2) 2/8 (3) 2/9
(4) 10/12 (5) 5/6 (6) 5/5
(7) 4/9 (8) 7/8 (9) 3/10
(10) 3/10 (11) 2/4 (12) 7/4
(13) 8/13 (14) 6/10 (15) 2/12
(16) 6/8 (17) 6/9 (18) 5/3
(19) 10/14 (20) 2/3 (21) 6/8
(22) 4/7 (23) 6/10 (24) 4/10
(25) 6/9 (26) 4/5 (27) 3/12
(28) 8/10 (29) 2/6 (30) 2/6
(31) 2/13 (32) 8/10 (33) 5/7
(34) 10/12 (35) 7/6 (36) 7/6
(37) 4/6 (38) 2/2 (39) 3/3
(40) 6/12 (41) 6/5 (42) 2/12
(43) 8/8 (44) 7/3 (45) 5/4

DAY 18
(1) 2/5 (2) 2/12 (3) 10/8
(4) 10/10 (5) 8/9 (6) 4/7
(7) 4/12 (8) 8/10 (9) 8/8
(10) 6/13 (11) 2/10 (12) 5/12
(13) 8/9 (14) 9/7 (15) 3/6
(16) 2/12 (17) 4/8 (18) 9/9
(19) 10/8 (20) 2/6 (21) 2/12
(22) 2/8 (23) 8/12 (24) 8/8
(25) 6/12 (26) 7/7 (27) 5/6
(28) 8/15 (29) 2/10 (30) 5/3
(31) 2/12 (32) 7/10 (33) 4/4
(34) 10/13 (35) 4/9 (36) 9/9
(37) 4/7 (38) 3/10 (39) 5/10
(40) 6/8 (41) 9/7 (42) 3/12
(43) 8/13 (44) 5/12 (45) 8/8

DAY 19
(1) 2/9 (2) 2/4 (3) 5/5
(4) 10/12 (5) 7/8 (6) 5/10
(7) 4/6 (8) 6/5 (9) 5/10
(10) 6/9 (11) 2/3 (12) 5/12
(13) 8/12 (14) 7/9 (15) 6/8
(16) 2/8 (17) 6/6 (18) 4/5
(19) 10/10 (20) 3/7 (21) 2/9
(22) 4/5 (23) 9/8 (24) 6/6
(25) 6/7 (26) 5/12 (27) 7/6
(28) 8/10 (29) 3/8 (30) 2/12
(31) 2/9 (32) 5/7 (33) 6/4
(34) 10/12 (35) 6/6 (36) 6/7
(37) 4/8 (38) 4/3 (39) 6/6
(40) 6/12 (41) 7/8 (42) 6/12
(43) 8/6 (44) 6/5 (45) 4/3

DAY 20
(1) 2/7 (2) 5/2 (3) 4/5
(4) 10/10 (5) 9/7 (6) 6/5
(7) 4/5 (8) 5/8 (9) 6/10
(10) 6/8 (11) 4/7 (12) 3/9
(13) 8/12 (14) 7/6 (15) 4/7
(16) 2/12 (17) 5/5 (18) 2/12
(19) 4/7 (20) 4/8 (21) 6/7
(22) 8/6 (23) 9/12 (24) 4/6
(25) 8/9 (26) 5/6 (27) 6/3
(28) 6/6 (29) 4/5 (30) 7/3
(31) 10/12 (32) 8/8 (33) 4/6
(34) 4/3 (35) 6/8 (36) 10/10
(37) 6/4 (38) 4/6 (39) 5/8
(40) 8/5 (41) 9/4 (42) 4/4
(43) 2/12 (44) 5/10 (45) 5/4

DAY 21
(1) 10/8 (2) 3/8 (3) 5/12
(4) 4/9 (5) 6/9 (6) 6/8
(7) 6/10 (8) 9/9 (9) 6/9
(10) 8/7 (11) 3/12 (12) 5/5
(13) 2/9 (14) 9/12 (15) 6/9
(16) 10/15 (17) 5/9 (18) 6/5
(19) 4/4 (20) 3/10 (21) 2/12
(22) 6/7 (23) 9/8 (24) 4/12
(25) 8/10 (26) 7/7 (27) 3/12
(28) 2/12 (29) 7/8 (30) 5/3
(31) 10/13 (32) 4/4 (33) 6/3
(34) 4/6 (35) 7/6 (36) 3/7
(37) 6/9 (38) 6/7 (39) 5/5
(40) 8/5 (41) 2/9 (42) 4/10
(43) 2/5 (44) 5/10 (45) 5/6

DAY 22
(1) 10/7 (2) 5/12 (3) 6/6
(4) 4/8 (5) 3/12 (6) 3/4
(7) 6/12 (8) 9/8 (9) 9/8
(10) 8/6 (11) 6/6 (12) 6/7
(13) 2/7 (14) 3/3 (15) 6/10
(16) 10/10 (17) 8/8 (18) 8/8
(19) 4/5 (20) 5/4 (21) 4/8
(22) 6/8 (23) 5/5 (24) 7/7
(25) 8/12 (26) 8/9 (27) 5/12
(28) 2/4 (29) 6/10 (30) 4/12
(31) 10/8 (32) 4/8 (33) 4/9
(34) 4/7 (35) 7/7 (36) 7/7
(37) 6/10 (38) 6/12 (39) 4/12
(40) 8/13 (41) 4/6 (42) 5/5
(43) 2/8 (44) 9/12 (45) 7/8

DAY 23
(1) 10/12 (2) 4/7 (3) 6/6
(4) 4/9 (5) 9/10 (6) 7/12
(7) 6/12 (8) 2/12 (9) 6/10
(10) 8/8 (11) 8/9 (12) 9/8
(13) 2/6 (14) 5/5 (15) 5/7
(16) 10/9 (17) 3/8 (18) 8/8
(19) 4/6 (20) 8/7 (21) 6/6
(22) 6/7 (23) 6/8 (24) 5/10
(25) 8/10 (26) 6/3 (27) 5/5
(28) 2/9 (29) 5/4 (30) 5/6
(31) 10/12 (32) 9/9 (33) 6/12
(34) 4/8 (35) 10/8 (36) 5/12
(37) 6/12 (38) 5/4 (39) 7/9
(40) 6/7 (41) 4/7 (42) 3/7
(43) 2/7 (44) 4/2 (45) 5/6

DAY 24
(1) 10/10 (2) 7/8 (3) 7/12
(4) 4/5 (5) 2/8 (6) 6/7
(7) 6/8 (8) 8/12 (9) 4/10
(10) 8/12 (11) 4/3 (12) 5/6
(13) 2/12 (14) 5/8 (15) 6/8
(16) 10/13 (17) 6/6 (18) 5/7
(19) 2/2 (20) 6/8 (21) 5/12
(22) 4/4 (23) 12/9 (24) 5/12
(25) 6/6 (26) 5/6 (27) 4/6
(28) 8/8 (29) 7/6 (30) 5/10
(31) 10/10 (32) 3/3 (33) 4/7
(34) 4/3 (35) 8/7 (36) 5/12
(37) 6/9 (38) 4/7 (39) 5/12
(40) 8/10 (41) 7/8 (42) 6/12
(43) 2/8 (44) 6/4 (45) 7/7

DAY 25
(1) 10/15 (2) 5/9 (3) 4/9
(4) 4/5 (5) 6/10 (6) 6/8
(7) 7/7 (8) 7/8 (9) 7/6
(10) 8/9 (11) 4/6 (12) 4/12
(13) 2/3 (14) 6/8 (15) 7/8
(16) 10/8 (17) 6/6 (18) 4/10
(19) 4/4 (20) 2/7 (21) 5/7
(22) 6/5 (23) 4/5 (24) 5/12
(25) 7/7 (26) 6/9 (27) 5/5
(28) 2/10 (29) 5/5 (30) 7/7
(31) 10/13 (32) 7/7 (33) 8/4
(34) 6/7 (35) 10/6 (36) 6/12
(37) 8/7 (38) 8/7 (39) 5/12
(40) 8/7 (41) 4/10 (42) 3/12
(43) 2/12 (44) 7/12 (45) 6/5

Subtracting Fractions - Answer Key (25 Days)

DAY 1
(1) 1/7	(2) 2/7	(3) 1/4
(4) 5/8	(5) 2/9	(6) 1/6
(7) 5/10	(8) 2/12	(9) 1/9
(10) 1/12	(11) 2/8	(12) 3/10
(13) 7/5	(14) 4/8	(15) 1/3
(16) 1/9	(17) 1/8	(18) 1/3
(19) 6/9	(20) 3/8	(21) 4/10
(22) 2/12	(23) 1/9	(24) 3/8
(25) 2/6	(26) 2/3	(27) 1/7
(28) 5/12	(29) 1/4	(30) 2/4
(31) 2/10	(32) 2/5	(33) 1/5
(34) 4/12	(35) 2/7	(36) 2/7
(37) 4/7	(38) 4/6	(39) 2/8
(40) 1/8	(41) 3/7	(42) 3/3
(43) 6/7	(44) 5/12	(45) 1/6

DAY 2
(1) 1/6	(2) 1/4	(3) 1/4
(4) 4/10	(5) 6/8	(6) 5/8
(7) 5/12	(8) 5/8	(9) 4/9
(10) 5/9	(11) 4/9	(12) 2/10
(13) 5/4	(14) 1/5	(15) 2/5
(16) 3/8	(17) 2/10	(18) 2/6
(19) 6/5	(20) 1/3	(21) 3/8
(22) 3/6	(23) 3/7	(24) 3/2
(25) 1/7	(26) 1/9	(27) 3/12
(28) 5/10	(29) 1/5	(30) 1/7
(31) 3/12	(32) 6/12	(33) 2/6
(34) 7/7	(35) 6/10	(36) 2/10
(37) 3/9	(38) 1/6	(39) 1/3
(40) 3/12	(41) 2/4	(42) 3/9
(43) 7/4	(44) 2/9	(45) 1/8

DAY 3
(1) 3/9	(2) 1/3	(3) 2/10
(4) 5/6	(5) 2/7	(6) 2/4
(7) 5/8	(8) 1/8	(9) 2/8
(10) 2/10	(11) 2/8	(12) 4/10
(13) 6/5	(14) 7/8	(15) 1/7
(16) 3/6	(17) 4/5	(18) 2/6
(19) 2/4	(20) 2/5	(21) 1/12
(22) 1/8	(23) 5/12	(24) 1/9
(25) 1/10	(26) 1/6	(27) 1/5
(28) 2/9	(29) 3/10	(30) 5/7
(31) 4/7	(32) 1/5	(33) 1/8
(34) 2/8	(35) 3/8	(36) 2/10
(37) 1/8	(38) 5/9	(39) 2/4
(40) 4/12	(41) 2/4	(42) 2/9
(43) 2/12	(44) 3/10	(45) 1/10

DAY 4
(1) 1/5	(2) 2/12	(3) 1/6
(4) 4/9	(5) 3/3	(6) 3/5
(7) 3/7	(8) 3/7	(9) 2/12
(10) 3/12	(11) 6/8	(12) 3/7
(13) 1/4	(14) 2/9	(15) 2/9
(16) 1/9	(17) 2/9	(18) 1/10
(19) 2/6	(20) 2/8	(21) 1/10
(22) 1/10	(23) 5/10	(24) 2/7
(25) 3/5	(26) 9/12	(27) 1/3
(28) 2/8	(29) 1/8	(30) 1/4
(31) 3/7	(32) 6/6	(33) 1/8
(34) 2/12	(35) 2/3	(36) 2/7
(37) 2/9	(38) 6/10	(39) 1/10
(40) 4/8	(41) 2/7	(42) 1/5
(43) 3/3	(44) 5/12	(45) 3/9

DAY 5
(1) 2/5	(2) 2/8	(3) 1/6
(4) 1/10	(5) 5/4	(6) 2/9
(7) 4/3	(8) 1/7	(9) 3/12
(10) 3/9	(11) 2/6	(12) 2/2
(13) 2/6	(14) 7/10	(15) 4/8
(16) 2/12	(17) 3/10	(18) 3/6
(19) 2/12	(20) 4/12	(21) 1/7
(22) 1/8	(23) 2/6	(24) 2/4
(25) 4/4	(26) 4/4	(27) 2/4
(28) 2/7	(29) 1/5	(30) 3/9
(31) 1/9	(32) 6/9	(33) 2/7
(34) 3/12	(35) 10/12	(36) 1/3
(37) 2/12	(38) 1/8	(39) 1/10
(40) 4/5	(41) 8/9	(42) 2/5
(43) 1/7	(44) 6/5	(45) 2/12

DAY 6
(1) 4/12	(2) 7/9	(3) 1/6
(4) 2/6	(5) 2/8	(6) 1/10
(7) 3/4	(8) 1/12	(9) 1/5
(10) 2/10	(11) 4/5	(12) 2/8
(13) 3/8	(14) 4/4	(15) 1/7
(16) 2/5	(17) 6/8	(18) 4/8
(19) 4/10	(20) 1/4	(21) 3/9
(22) 1/12	(23) 3/5	(24) 4/7
(25) 2/7	(26) 9/10	(27) 2/4
(28) 2/3	(29) 1/6	(30) 1/7
(31) 1/5	(32) 3/7	(33) 2/4
(34) 3/12	(35) 3/7	(36) 1/6
(37) 1/8	(38) 3/9	(39) 1/10
(40) 3/12	(41) 6/8	(42) 2/6
(43) 3/4	(44) 1/6	(45) 3/8

DAY 7
(1) 2/10	(2) 3/10	(3) 5/9
(4) 4/6	(5) 2/10	(6) 6/10
(7) 1/9	(8) 5/10	(9) 2/10
(10) 3/8	(11) 2/12	(12) 2/7
(13) 2/5	(14) 1/10	(15) 1/3
(16) 6/8	(17) 2/12	(18) 1/3
(19) 4/10	(20) 3/7	(21) 2/12
(22) 1/5	(23) 2/8	(24) 3/9
(25) 6/9	(26) 6/10	(27) 1/10
(28) 1/12	(29) 1/5	(30) 1/7
(31) 5/8	(32) 7/9	(33) 1/9
(34) 4/6	(35) 1/3	(36) 1/10
(37) 4/10	(38) 2/5	(39) 1/4
(40) 3/7	(41) 8/10	(42) 1/5
(43) 4/8	(44) 4/8	(45) 1/8

DAY 8
(1) 3/12	(2) 1/3	(3) 1/10
(4) 3/5	(5) 1/7	(6) 2/7
(7) 2/4	(8) 2/12	(9) 2/10
(10) 3/12	(11) 2/8	(12) 1/5
(13) 3/7	(14) 2/4	(15) 2/12
(16) 1/6	(17) 2/4	(18) 3/9
(19) 1/10	(20) 2/6	(21) 2/10
(22) 3/12	(23) 2/2	(24) 3/4
(25) 4/4	(26) 1/5	(27) 3/7
(28) 5/6	(29) 4/6	(30) 2/2
(31) 1/7	(32) 5/12	(33) 3/10
(34) 2/9	(35) 1/7	(36) 6/8
(37) 2/4	(38) 1/9	(39) 6/12
(40) 2/5	(41) 8/8	(42) 1/3
(43) 1/6	(44) 1/10	(45) 1/8

DAY 9
(1) 2/10	(2) 9/10	(3) 4/6
(4) 3/8	(5) 2/7	(6) 4/10
(7) 3/12	(8) 5/10	(9) 2/9
(10) 2/8	(11) 8/10	(12) 1/5
(13) 1/12	(14) 2/5	(15) 3/6
(16) 2/7	(17) 8/9	(18) 1/3
(19) 3/9	(20) 2/3	(21) 1/7
(22) 4/10	(23) 1/8	(24) 2/4
(25) 2/6	(26) 3/5	(27) 3/8
(28) 2/12	(29) 3/4	(30) 2/4
(31) 1/7	(32) 6/12	(33) 2/9
(34) 2/8	(35) 2/7	(36) 2/10
(37) 3/6	(38) 3/12	(39) 1/10
(40) 2/12	(41) 3/3	(42) 2/6
(43) 3/4	(44) 1/4	(45) 2/8

DAY 10
(1) 2/12	(2) 5/6	(3) 3/6
(4) 4/7	(5) 2/5	(6) 3/7
(7) 2/6	(8) 7/10	(9) 3/5
(10) 1/9	(11) 5/8	(12) 2/9
(13) 2/10	(14) 4/7	(15) 3/5
(16) 3/5	(17) 1/12	(18) 3/7
(19) 2/8	(20) 4/9	(21) 4/10
(22) 2/5	(23) 9/8	(24) 2/12
(25) 2/7	(26) 2/10	(27) 2/10
(28) 2/12	(29) 6/9	(30) 2/6
(31) 1/4	(32) 3/3	(33) 2/7
(34) 1/9	(35) 1/3	(36) 6/9
(37) 4/10	(38) 1/5	(39) 2/12
(40) 5/7	(41) 4/6	(42) 2/10
(43) 1/8	(44) 2/4	(45) 4/10

DAY 11
(1) 2/10	(2) 3/7	(3) 1/8
(4) 1/6	(5) 3/9	(6) 4/8
(7) 3/10	(8) 2/8	(9) 1/9
(10) 2/7	(11) 1/10	(12) 2/3
(13) 3/8	(14) 3/12	(15) 1/4
(16) 2/6	(17) 2/10	(18) 2/5
(19) 1/12	(20) 3/9	(21) 2/7
(22) 6/4	(23) 4/8	(24) 3/6
(25) 3/7	(26) 2/10	(27) 3/7
(28) 2/5	(29) 1/6	(30) 1/12
(31) 5/9	(32) 2/7	(33) 1/4
(34) 1/10	(35) 4/10	(36) 5/10
(37) 4/7	(38) 1/5	(39) 3/8
(40) 2/6	(41) 4/12	(42) 4/9
(43) 1/8	(44) 1/7	(45) 1/5

DAY 12
(1) 1/9	(2) 1/10	(3) 1/10
(4) 1/10	(5) 5/8	(6) 1/3
(7) 1/5	(8) 3/9	(9) 3/7
(10) 1/8	(11) 3/7	(12) 1/9
(13) 1/10	(14) 3/5	(15) 1/6
(16) 2/10	(17) 1/8	(18) 2/12
(19) 1/4	(20) 3/10	(21) 2/10
(22) 2/8	(23) 2/10	(24) 1/7
(25) 1/6	(26) 3/6	(27) 2/6
(28) 1/7	(29) 1/9	(30) 2/3
(31) 1/9	(32) 3/5	(33) 1/8
(34) 2/10	(35) 1/10	(36) 2/5
(37) 2/7	(38) 1/10	(39) 3/2
(40) 1/6	(41) 1/7	(42) 1/7
(43) 1/12	(44) 2/12	(45) 1/10

Subtracting Fractions - Answer Key

DAY 13
(1) 1/10 (2) 1/10 (3) 2/7
(4) 1/10 (5) 4/8 (6) 3/6
(7) 1/7 (8) 3/12 (9) 1/5
(10) 1/3 (11) 1/4 (12) 2/8
(13) 1/12 (14) 1/10 (15) 1/12
(16) 1/8 (17) 1/6 (18) 1/6
(19) 1/4 (20) 1/3 (21) 2/5
(22) 1/5 (23) 2/4 (24) 2/10
(25) 1/10 (26) 2/8 (27) 1/8
(28) 1/9 (29) 2/9 (30) 1/6
(31) 1/2 (32) 1/5 (33) 1/9
(34) 1/6 (35) 3/6 (36) 1/10
(37) 1/12 (38) 4/10 (39) 1/12
(40) 2/10 (41) 1/12 (42) 3/9
(43) 3/6 (44) 2/6 (45) 3/4

DAY 14
(1) 5/7 (2) 1/8 (3) 1/12
(4) 1/3 (5) 1/7 (6) 1/8
(7) 2/9 (8) 2/12 (9) 1/9
(10) 2/5 (11) 1/4 (12) 1/10
(13) 2/10 (14) 3/9 (15) 1/12
(16) 4/7 (17) 1/10 (18) 1/7
(19) 2/8 (20) 4/3 (21) 3/5
(22) 2/6 (23) 2/8 (24) 2/4
(25) 2/5 (26) 1/9 (27) 1/6
(28) 3/10 (29) 3/10 (30) 2/7
(31) 2/4 (32) 2/10 (33) 2/9
(34) 2/9 (35) 3/2 (36) 1/10
(37) 2/6 (38) 2/10 (39) 1/12
(40) 2/9 (41) 1/8 (42) 1/7
(43) 2/7 (44) 2/4 (45) 1/12

DAY 15
(1) 2/4 (2) 1/10 (3) 2/6
(4) 2/12 (5) 1/9 (6) 1/3
(7) 3/8 (8) 2/7 (9) 3/8
(10) 2/10 (11) 1/4 (12) 1/4
(13) 2/10 (14) 1/8 (15) 1/5
(16) 1/8 (17) 1/7 (18) 1/9
(19) 3/5 (20) 3/6 (21) 2/10
(22) 2/7 (23) 6/3 (24) 2/12
(25) 1/10 (26) 2/7 (27) 2/7
(28) 2/9 (29) 2/9 (30) 2/8
(31) 1/3 (32) 1/7 (33) 2/6
(34) 2/9 (35) 1/8 (36) 1/8
(37) 8/10 (38) 1/12 (39) 1/7
(40) 2/7 (41) 1/6 (42) 1/10
(43) 2/6 (44) 2/10 (45) 9/12

DAY 16
(1) 1/5 (2) 7/8 (3) 2/9
(4) 1/8 (5) 1/12 (6) 1/5
(7) 3/4 (8) 3/9 (9) 1/8
(10) 2/12 (11) 1/4 (12) 5/7
(13) 3/10 (14) 1/6 (15) 3/12
(16) 4/8 (17) 1/7 (18) 2/3
(19) 8/7 (20) 5/10 (21) 3/4
(22) 1/5 (23) 1/8 (24) 3/5
(25) 2/10 (26) 2/5 (27) 1/7
(28) 5/6 (29) 8/7 (30) 4/12
(31) 5/9 (32) 1/9 (33) 6/10
(34) 1/7 (35) 1/12 (36) 3/9
(37) 4/10 (38) 1/12 (39) 3/4
(40) 1/6 (41) 1/12 (42) 5/8
(43) 2/12 (44) 1/12 (45) 2/12

DAY 17
(1) 8/8 (2) 4/9 (3) 3/7
(4) 1/6 (5) 1/5 (6) 3/12
(7) 3/8 (8) 1/10 (9) 3/9
(10) 5/7 (11) 1/4 (12) 1/10
(13) 4/10 (14) 6/12 (15) 2/13
(16) 2/9 (17) 1/3 (18) 4/8
(19) 5/3 (20) 2/8 (21) 2/14
(22) 6/10 (23) 1/10 (24) 3/7
(25) 2/5 (26) 1/12 (27) 4/9
(28) 6/9 (29) 7/7 (30) 5/10
(31) 4/10 (32) 1/7 (33) 6/13
(34) 3/6 (35) 1/12 (36) 1/12
(37) 8/2 (38) 1/3 (39) 6/6
(40) 2/5 (41) 4/12 (42) 2/12
(43) 3/3 (44) 1/4 (45) 4/8

DAY 18
(1) 7/12 (2) 7/8 (3) 5/5
(4) 2/9 (5) 2/7 (6) 2/10
(7) 7/10 (8) 2/8 (9) 3/12
(10) 6/10 (11) 1/12 (12) 5/13
(13) 5/7 (14) 1/6 (15) 1/9
(16) 4/8 (17) 5/7 (18) 6/12
(19) 4/6 (20) 3/12 (21) 1/8
(22) 4/12 (23) 6/3 (24) 5/8
(25) 3/7 (26) 1/6 (27) 2/12
(28) 8/5 (29) 1/3 (30) 4/15
(31) 1/10 (32) 1/4 (33) 3/12
(34) 3/4 (35) 7/7 (36) 1/13
(37) 1/10 (38) 3/10 (39) 6/7
(40) 1/7 (41) 1/6 (42) 2/10
(43) 1/12 (44) 4/8 (45) 4/13

DAY 19
(1) 7/4 (2) 3/5 (3) 5/9
(4) 1/8 (5) 3/10 (6) 4/12
(7) 2/5 (8) 1/10 (9) 6/6
(10) 6/3 (11) 1/12 (12) 1/9
(13) 1/9 (14) 4/8 (15) 5/12
(16) 1/6 (17) 2/5 (18) 2/3
(19) 1/7 (20) 2/9 (21) 1/10
(22) 1/8 (23) 2/6 (24) 3/5
(25) 1/12 (26) 3/6 (27) 5/7
(28) 1/8 (29) 4/12 (30) 3/10
(31) 1/7 (32) 4/4 (33) 8/9
(34) 2/6 (35) 3/7 (36) 2/12
(37) 2/3 (38) 2/6 (39) 4/4
(40) 1/8 (41) 4/12 (42) 2/12
(43) 2/5 (44) 2/3 (45) 4/6

DAY 20
(1) 3/2 (2) 2/5 (3) 4/7
(4) 1/7 (5) 3/5 (6) 2/10
(7) 1/8 (8) 4/10 (9) 3/5
(10) 2/7 (11) 1/9 (12) 5/8
(13) 3/6 (14) 4/7 (15) 2/12
(16) 1/5 (17) 7/12 (18) 8/12
(19) 1/8 (20) 4/7 (21) 4/7
(22) 1/12 (23) 4/4 (24) 4/8
(25) 1/6 (26) 4/3 (27) 4/9
(28) 2/5 (29) 3/3 (30) 5/6
(31) 2/8 (32) 2/6 (33) 5/12
(34) 1/8 (35) 1/10 (36) 1/3
(37) 1/6 (38) 1/8 (39) 2/4
(40) 1/9 (41) 7/4 (42) 4/5
(43) 1/10 (44) 3/4 (45) 5/12

DAY 21
(1) 1/8 (2) 1/12 (3) 4/8
(4) 3/9 (5) 3/8 (6) 5/9
(7) 3/4 (8) 2/9 (9) 2/10
(10) 1/12 (11) 1/7 (12) 4/7
(13) 1/8 (14) 2/9 (15) 5/9
(16) 1/9 (17) 4/5 (18) 4/15
(19) 1/10 (20) 1/12 (21) 5/4
(22) 1/8 (23) 5/12 (24) 2/7
(25) 1/7 (26) 1/12 (27) 2/10
(28) 3/5 (29) 3/3 (30) 6/12
(31) 2/4 (32) 2/3 (33) 1/13
(34) 1/6 (35) 1/7 (36) 6/7
(37) 2/7 (38) 1/5 (39) 3/7
(40) 7/9 (41) 2/10 (42) 1/6
(43) 1/10 (44) 3/6 (45) 8/13

DAY 22
(1) 1/12 (2) 2/6 (3) 2/7
(4) 1/7 (5) 1/4 (6) 3/12
(7) 1/8 (8) 1/9 (9) 4/12
(10) 2/6 (11) 2/7 (12) 4/6
(13) 1/3 (14) 2/10 (15) 5/7
(16) 1/8 (17) 3/8 (18) 3/10
(19) 1/4 (20) 2/8 (21) 5/5
(22) 8/5 (23) 5/7 (24) 3/8
(25) 2/9 (26) 1/12 (27) 5/12
(28) 2/10 (29) 2/12 (30) 4/4
(31) 2/8 (32) 2/9 (33) 1/8
(34) 5/7 (35) 3/7 (36) 2/7
(37) 2/8 (38) 2/12 (39) 2/10
(40) 2/6 (41) 2/5 (42) 5/13
(43) 1/12 (44) 3/8 (45) 5/8

DAY 23
(1) 5/7 (2) 4/6 (3) 1/12
(4) 1/10 (5) 3/12 (6) 6/9
(7) 5/12 (8) 2/10 (9) 4/12
(10) 2/9 (11) 3/9 (12) 2/8
(13) 1/5 (14) 3/7 (15) 7/6
(16) 1/8 (17) 2/8 (18) 1/9
(19) 3/7 (20) 2/6 (21) 2/6
(22) 6/8 (23) 3/10 (24) 5/7
(25) 2/3 (26) 1/5 (27) 2/10
(28) 3/4 (29) 3/9 (30) 4/9
(31) 3/9 (32) 4/12 (33) 2/12
(34) 2/7 (35) 1/12 (36) 3/8
(37) 3/4 (38) 3/9 (39) 4/12
(40) 4/7 (41) 1/8 (42) 4/6
(43) 2/2 (44) 1/6 (45) 4/7

DAY 24
(1) 3/10 (2) 1/12 (3) 2/10
(4) 6/8 (5) 2/7 (6) 3/5
(7) 2/12 (8) 2/10 (9) 1/8
(10) 4/3 (11) 3/6 (12) 2/12
(13) 1/8 (14) 2/9 (15) 3/12
(16) 4/6 (17) 3/7 (18) 2/13
(19) 4/8 (20) 1/12 (21) 4/2
(22) 2/9 (23) 3/12 (24) 6/4
(25) 1/5 (26) 2/7 (27) 5/6
(28) 3/6 (29) 1/10 (30) 2/8
(31) 1/3 (32) 2/10 (33) 5/10
(34) 3/7 (35) 3/5 (36) 3/3
(37) 2/4 (38) 2/12 (39) 3/10
(40) 3/8 (41) 1/12 (42) 4/10
(43) 2/4 (44) 1/7 (45) 4/8

DAY 25
(1) 3/9 (2) 3/9 (3) 2/15 (4) 4/10 (5) 4/8
(6) 3/5 (7) 1/8 (8) 3/6 (9) 3/7 (10) 2/6
(11) 2/12 (12) 4/9 (13) 2/8 (14) 1/8 (15) 5/3
(16) 1/6 (17) 2/10 (18) 3/8 (19) 4/7 (20) 1/7
(21) 6/4 (22) 4/5 (23) 3/12 (24) 2/5 (25) 2/9
(26) 1/5 (27) 3/7 (28) 3/5 (29) 1/9 (30) 4/10
(31) 3/7 (32) 1/8 (33) 13/4 (34) 10/4 (35) 2/12
(36) 3/6 (37) 2/12 (38) 1/7 (39) 2/8 (40) 2/10
(41) 9/1 (42) 3/12 (43) 6/2 (44) 5/7 (45) 7/12

Fractions Chart: Parts of a Whole, Decimals, & Percentages

Fraction		Decimal	Percent	Name	Bar model
1		1.0	100%	One whole	1
$\frac{1}{2}$		0.5	50%	One half	$\frac{1}{2}$ $\frac{1}{2}$
$\frac{1}{3}$		0.333	33.3%	One third	$\frac{1}{3}$ $\frac{1}{3}$ $\frac{1}{3}$
$\frac{1}{4}$		0.25	25%	One quarter	$\frac{1}{4}$ $\frac{1}{4}$ $\frac{1}{4}$ $\frac{1}{4}$
$\frac{1}{5}$		0.20	20%	One fifth	$\frac{1}{5}$ $\frac{1}{5}$ $\frac{1}{5}$ $\frac{1}{5}$ $\frac{1}{5}$
$\frac{1}{6}$		0.166	16.6%	One sixth	$\frac{1}{6}$ $\frac{1}{6}$ $\frac{1}{6}$ $\frac{1}{6}$ $\frac{1}{6}$ $\frac{1}{6}$
$\frac{1}{7}$		0.143	14.29%	One seventh	$\frac{1}{7}$ $\frac{1}{7}$ $\frac{1}{7}$ $\frac{1}{7}$ $\frac{1}{7}$ $\frac{1}{7}$ $\frac{1}{7}$
$\frac{1}{8}$		0.125	12.5%	One eighth	$\frac{1}{8}$ $\frac{1}{8}$ $\frac{1}{8}$ $\frac{1}{8}$ $\frac{1}{8}$ $\frac{1}{8}$ $\frac{1}{8}$ $\frac{1}{8}$
$\frac{1}{9}$		0.1111	11.11%	One ninth	$\frac{1}{9}$ $\frac{1}{9}$ $\frac{1}{9}$ $\frac{1}{9}$ $\frac{1}{9}$ $\frac{1}{9}$ $\frac{1}{9}$ $\frac{1}{9}$ $\frac{1}{9}$
$\frac{1}{10}$		0.10	10%	One tenth	$\frac{1}{10}$ $\frac{1}{10}$ $\frac{1}{10}$ $\frac{1}{10}$ $\frac{1}{10}$ $\frac{1}{10}$ $\frac{1}{10}$ $\frac{1}{10}$ $\frac{1}{10}$ $\frac{1}{10}$
$\frac{1}{11}$		0.0909	9.09%	One eleventh	$\frac{1}{11}$ $\frac{1}{11}$ $\frac{1}{11}$ $\frac{1}{11}$ $\frac{1}{11}$ $\frac{1}{11}$ $\frac{1}{11}$ $\frac{1}{11}$ $\frac{1}{11}$ $\frac{1}{11}$ $\frac{1}{11}$
$\frac{1}{12}$		0.0833	8.33%	One twelfth	$\frac{1}{12}$ $\frac{1}{12}$ $\frac{1}{12}$ $\frac{1}{12}$ $\frac{1}{12}$ $\frac{1}{12}$ $\frac{1}{12}$ $\frac{1}{12}$ $\frac{1}{12}$ $\frac{1}{12}$ $\frac{1}{12}$ $\frac{1}{12}$

ISBN: 9798887200125

ISBN: 9798887200217

ISBN: 9798887200170

Hey there, why not take a look at some of our other books? We've got a great selection!

abcZbook Press

Certificate of Excellence Award in

Fractions
Beginner Level

Congratulations!

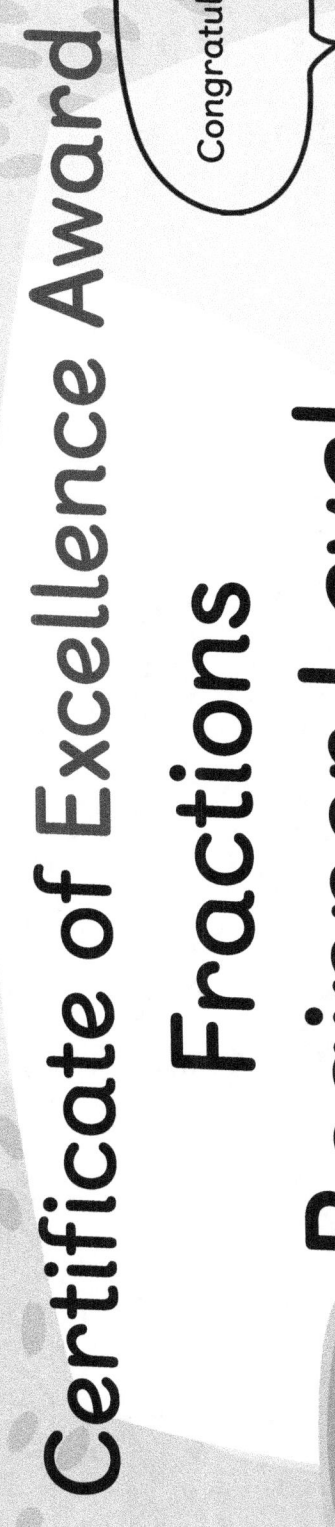

By:

Date:

You are a Star!

abcZbook